OUR PRESENT COMPLAINT

OUR PRESENT COMPLAINT

American Medicine, Then and Now

CHARLES E. ROSENBERG

THE JOHNS HOPKINS UNIVERSITY PRESS
Baltimore

The Johns Hopkins University Press
2715 North Charles Street
Baltimore, Maryland 21218-4363
www.press.jhu.edu

LIBRARY OF CONGRESS CATALOGING-IN-PUBLICATION DATA

Rosenberg, Charles E.
    Our present complaint : American medicine, then and now / Charles E. Rosenberg
       p. ; cm.
    Includes bibliographical references and index
    ISBN-13: 978-0-8018-8715-4 (hardcover : alk. paper)
    ISBN-13: 978-0-8018-8716-1 (pbk. : alk. paper)
    ISBN-10: 0-8018-8715-1 (hardcover : alk. paper)
    ISBN-10: 0-8018-8716-X (pbk. : alk. paper)
     1. Medicine—United States—History. 2. Medical care—United States—History. I. Title.
     [DNLM: 1. Patient Satisfaction—United States. 2. Delivery of Health Care—United
States. 3. Health Policy—United States. 4. Physician's Role—United States. W 85 R813o
2007]
R151.R67 2007
610—dc22   2007013960

A catalog record for this book is available from the British Library.

*Special discounts are available for bulk purchases of this book. For more information,
please contact Special Sales at 410-516-6936 or specialsales@press.jhu.edu.*

CONTENTS

- -

CHAPTER 1.

Introduction: *The History of Our Present Complaint*   1

CHAPTER 2.

The Tyranny of Diagnosis: *Specific Entities and Individual Experience*   13

CHAPTER 3.

Contested Boundaries: *Psychiatry, Disease, and Diagnosis*   38

CHAPTER 4.

Banishing Risk: *Or, the More Things Change, the More They Remain the Same*   60

CHAPTER 5.

Pathologies of Progress: *The Idea of Civilization as Risk*   77

CHAPTER 6.

The New Enchantment: *Genetics, Medicine, and Society*   96

CHAPTER 7.

Alternative to What? Complementary to Whom? *On the Scientific Project in Medicine* 113

CHAPTER 8.

Holism in Twentieth-Century Medicine: *Always in Opposition* 139

CHAPTER 9.

Mechanism and Morality: *On Bioethics in Context* 166

CHAPTER 10.

Anticipated Consequences: *Historians, History, and Health Policy* 185

Acknowledgments 207
Index 209

OUR PRESENT COMPLAINT

- -

CHAPTER 1

INTRODUCTION

The History of Our Present Complaint

THESE ARE DIFFICULT TIMES IN WHICH TO practice medicine. These are the best of times, and these are the worst of times for American clinicians—and for their patients. Never have physicians been able to intervene more effectively in the body, never have they understood more about the mechanisms determining health and disease, yet never have practitioners felt themselves more constrained—if not besieged—by bureaucratic guidelines and intrusive administrative oversight. Increasing technical capacity seems wedded to diminishing autonomy, an odd situation indeed. Medicine has been technologically reinvented in the past half century, yet it remains in some ways what it has always been, an intensely personal effort to deal with the pain and incapacity of particular men and women. The patient is always an individual—even as she or he is defined by aggregate findings and generalized understandings—and it is the clinician who remains at the interface between our bureaucratic world of agreed-upon, if ever-changing, data, procedures, practices, and that individual.

The sick and the anxious well and their families find themselves in a parallel world of inconsistent expectation. And we are all—even physicians—patients or patients-in-waiting, members of patient families. Although we have never before expected so much from medical science, our technology-based hopes are balanced by a symmetrical ambivalence. Science-based technology promises lives enhanced and mortality forestalled yet threatens a dystopia of life artificially extended in pain and incapacity. There is also an economic price to be paid. Even the most optimistic advocate of innovation in medicine cannot ignore ever-increasing health-care costs, costs associated in some measure with that technology we so much admire. And, as we are equally well aware, access to clinical services is far from universal or equitable. As I write this introduction, more than forty million Americans lack health insurance and medical expenses remain a major cause of bankruptcy. Still another paradox complicates the relationship between society and medicine. Though expectations of therapeutic efficacy have never been more euphoric and patients appear to trust their own physicians, respect for the profession of medicine has declined.[1] The pharmaceutical industry is trusted even less.

Government, too, has come gradually to play an indispensable but ill-defined role in health care. Yet the state and the corporate world—and, in fact, the market itself—have become additional sources of anxiety, of hopes and unmet expectations. One need only pick up the morning newspaper or listen to a newscast to encounter stories of scientific "breakthroughs," on the one hand, and of a variety of dismayingly recurrent dysfunctions, on the other—aggressively marketed drugs with unanticipated side effects, regulatory conflicts of interest, inadequate or absent health care for many Americans, epidemics of ills such as asthma or adult-onset diabetes that seem linked ironically and structurally with aspects of economic growth and an altered physical environment. Progress does not come without cost. We also live with the apocalyptic threats of emergent diseases and bioterrorism, as well as the mundane irritations of an often rushed, impersonal, profit-oriented, fragmented, and technology-dependent clinical practice— and everywhere the specter of increasing costs in an aging population. It is a patient's, a physician's, and a policymaker's dilemma. Our present complaint of unrealistic medical expectations and everyday imperfection, of physicians and patients alike uncomfortable in their roles has produced a chronic attitudinal malaise enlivened by acute episodes of more sharply focused discomfort. This book represents a historian's attempt to put such contradictions into a longer-term perspective, to link past and present and suggest some ways of thinking about prospects for our new healing order, about our notions of disease and efficacy, and about the medical profession's peculiar role and responsibilities.

Every aspect of medicine has changed since our nation's founding, when Philadelphia was America's first capital and largest city and George Washington was the new nation's first president. Practice then ordinarily took place in the patient's home; family practice meant just that. A physician not only treated every member of the family—and often servants or apprentices—when they became ill but also cared for them in their own bedrooms. Hospital or almshouse care was a stigmatizing necessity for a minority among the urban poor in a still overwhelmingly rural population. With a handful of exceptions—some quacks and, in their different way, a few members of the urban consultant elite—there were no specialists.

Sickness itself was as variable as its sufferers. Disease was conceived of not as a set of specific entities, each with a characteristic and generally predictable course and underlying mechanism, but as a physiological state of the individual patient; all causation was multicausal (with a few exceptions, most conspicuously smallpox, which did seem to be spread by something specific and material). Even frightening epidemics, such as the yellow fever that visited Philadelphia in 1793, could be construed as the consequence of a peculiarly tainted microenvironment—presumably something in the atmosphere—coupled with an individual idiosyncrasy, which explained why some succumbed, some recovered, and others never fell ill during a local outbreak.[2] Clinical efficacy rested on the ability to recognize and manage such idiosyncrasy—and this meant not just an individual's biological attributes but also his or her social and emotional identity. Success in practice meant functioning as part of a small and interactive social world.

By the beginning of the twentieth century, all this had begun to change. Even respectable people might be treated in hospitals, which were no longer exclusively urban welfare institutions as they had been through almost all of the nineteenth century.[3] Specialists and specialty societies and journals had come into being and postgraduate clinical training had become a necessity for ambitious practitioners. Disease was now conceived of as a specific entity, defined by urine and blood tests and temperature curves as well as by the experienced clinicians' eyes and fingertips.[4] The germ theory and routinization of bacteriology in the last two decades of the nineteenth century seemed to both explain and legitimate this newly specific way of thinking about disease. These linked conceptual and institutional changes had implications for patient care, not all of them unambiguously positive. As early as the first decade of the twentieth century, would-be reformers of the medical profession warned that clinical practice was becoming impersonal and dehumanizing, especially in the hospital's rooms and wards. We were in danger of treating organs and diseases, not people; the kidney in Ward 4, not Mrs. Smith or Mr. Jones, such critics argued. Patients were in danger of becoming numbers on charts, shadows on x-ray plates, and smears on slides.

In retrospect such criticisms of a laboratory-oriented reductionist medicine seem no more than a distant early warning, in large part apparent only to the future-oriented and socially committed. The great bulk of medicine remained from today's perspective technologically primitive. Most clinical practice in 1900 was closer to that of 1800 than that of 2000. By the middle of the twentieth century, the pace of change had quickened dramatically. Physicians were increasingly able to intervene in the course of disease—and to base those interventions on ever-more-sophisticated (and capital-intensive) laboratory and imaging capacities. The first half of the twentieth century had given birth to a sequence of confidence-inspiring therapeutic innovations: insulin in the early 1920s, the sulfonamides and antibiotics in the 1930s and 1940s, steroids in the 1940s, and so on. These achievements were visible for anyone who could read a newspaper or magazine, even if they did not encounter them in their experience as patients. For the first time, medicine could—it seemed—routinely alter the course of life-threatening ailments or, as in the case of polio, avert them altogether. Science was transforming the physician's role and the patient's expectations.

Equally significant was the impact of an increasingly activist federal government. World War II dramatically increased the role of the public sector in medicine. Earlier precedents for federal government support of research—and even care, as in the Veterans' Administration—expanded dramatically after the war.[5] In hospital construction, in research, in veterans' programs, in clinical training, and then in the mid-1960s with Medicare and Medicaid, Washington began to play a substantive role in the delivery of health care. The interposition of third-party payers between physician and patient had begun even earlier. A variety of experiments in not-for-profit prepayment had preceded the invention of Blue Cross in the 1930s and by the 1950s Blue Cross and Blue Shield had become a normal part of America's medical environment. The West Coast's Kaiser Plan and a few other comprehensive schemes suggested an ever greater level of all-inclusive—and thus centrally administered—prepaid care.

Meanwhile in the years after the Second World War and—seemingly unrelated at the time to most observers—American medicine gradually came to accept the randomized clinical trial as what has come to be called the "gold standard" in evaluating therapeutics, a series of intellectual and administrative innovations given greater relevance by the thalidomide scandal and the subsequent accretion of proactive and preemptive power by the Food and Drug Administration. All of these events are familiar to historians, as they are to most Americans who lived through the 1950s and 1960s.[6]

Two fundamental and linked aspects of these events were—and are—less immediately visible. First was medicine's growing centralization and bureaucratization. Scale and institutional complexity inevitably brought the need for uniformity and control—not only in credentialing but also in the implicit role of government. Third-party payment would, over time, create pressures for increased control over participating physicians and hospitals. Ironically, the federal government's original lack of budget discipline in the 1960s has resulted in an ever-more-pressing need to moderate costs through managing care. We are still contending with increasing health expenditures in a technology—and, thus, capital-intensive—environment. Second, I would like to argue, was the way in which these changes in the evaluation of clinical evidence, in government policy, and in the public negotiation of diagnostic and treatment standards cumulatively helped shape a new way of thinking about and experiencing disease. Disease has become a bureaucratic—and, thus, social and administrative—as well as biological and conceptual—entity.

What do I mean when I describe disease as a "social entity"? I refer to a web of practice guidelines, disease protocols, laboratory and imaging results, meta-analyses, and consensus conferences. These practices and procedures have over time come to constitute a seemingly objective and inescapable framework of disease categories, a framework that increasingly specifies diagnostic criteria and dictates appropriate therapeutic choices. In America's peculiar hybrid health-care system, layers of hospital and managed care administrators enforce these disease-based guidelines. The past generation's revolution in information technology has only exacerbated and intensified these trends—in parallel with the research and marketing strategies of major pharmaceutical companies that have also turned on the diffusion and articulation of sometimes hypothetical disease entities. This web of complex relationships has created a new reality for practitioners and patients alike. Physicians have had their choices increasingly constrained—if, in some ways, enhanced. For the sick, such ways of conceptualizing and treating disease have come to constitute a tangible aspect of their illness experience.

Of course, every society has entertained ideas about disease and its treatment; patients have never been blank slates. Since classical antiquity, laypeople and practitioners have been influenced by theoretical notions about the nature of disease and its diagnosis, prognosis, and treatment, and the sick have inevitably been affected by such ideas and related practices. Think of the generations of sufferers who were bled, sweated, puked, or purged to balance their humors. But never has the infrastructure of ideas, practices, thresholds, and protocols that comes between agreed-upon knowledge and the individual patient been so tightly woven and bureaucratically crafted. Today's highly visible discussions surrounding "evidence-based medicine" demonstrate the centrality and intractability of such relationships and standardizing practices, at once components and rationale for the distribution of power and knowledge.

Yet, as I have emphasized, we are left with that inconveniently subjective object, the patient—in whose body these abstract entities manifest themselves. This is the characteristic split screen that faces today's clinician: the tension between illness in the individual patient and disease as crystallized and made socially real in the laboratory's and the epidemiologist's outputs and inscriptions, practice guidelines, and algorithms. To this body of data are added institutional pressures to create consistency-enforcing mechanisms ranging from consensus conferences to practice guidelines. Bedside, clinic, and physician's office are the points at which the mandates of best—and increasingly most economically rational—practice bump up against the unique reality of the individual patient and challenge the physician's traditional autonomy.

This is no simple problem with an easy resolution. It is difficult to weigh individual clinical decisions—how to treat a particular woman or man—against the collective good as articulated in practice guidelines and cost-benefit analyses. General truth is not always specific truth—at the level of technical decision or moral choice, and these dimensions of the same action cannot easily be disentangled. Even the most experienced clinicians have often to choose among a variety of uncertain options. And it is well documented that clinicians and institutions fail to consistently adhere to even well-established practice guidelines. Economic rationality may, and perhaps should, trump certain individual choices; evidence-based practice guidelines will inevitably lead to particular misjudgments at particular bedsides—but can save us from a larger number of missteps in other cases. How do we balance an optimum outcome for a particular individual with the greatest good for the greatest number?

This is not entirely a novel dilemma. Physicians have always had to balance the individual instance against a general consensus, and I should emphasize that these difficult choices are not limited to available diagnostic and therapeutic decisions; a physician's choices at the particular bedside are also constrained by more general—and, in a sense, prior—political and policy realities such as proper allocation of social resources and attitudes toward centralization and the state, toward race, gender, and age. Therapeutic decisions are never made in a disembodied intellectual space. We are all embedded in an ever-more-elaborate institutional and intellectual, governmental and corporate complex upon which we have come to depend and in which we—physicians and patients alike—have rested our collective hopes for medical deliverance from pain and premature mortality.

It engenders a feeling of paradox, the juxtaposition of a powerful faith in scientific medicine with a widespread discontent at the circumstances in which it is made available. It is a set of attitudes and expectations postmodern as well as quintessentially modern. We are well aware of the irrationalities and inequities structured into America's much-praised health-care system (or, more accurately praised for its advanced technical capacity, its ability to manage worst-case scenarios). Some of these inconsistencies are the consequences of America's political and ideological history; others are, as I have argued, built into the very fabric of the technical and highly specialized medicine so often admired, yet often criticized for its costs and its less-than-impressive results as compared to those of other industrialized countries that spend far less per capita for health care.

It is not only the costs of medical care and the relentless demographic pressures of an aging society that make us uncomfortable with contemporary health policies. It is more than status anxiety that makes physicians, as well as patients, uncomfortable with the contemporary realities of an increasingly bureaucratized and market-oriented medicine. The crisis we are experiencing is one of authority and control, as I have argued, as well as expectation and public policy; but it is also a crisis in values and orientations. It might be thought of as a crisis in maintaining the profession's appropriate and defining balance among a variety of not-always-consistent identities: medicine as humane caring, as applied science, as marketplace actor, and as an object of public policy.

Some of these paradoxes are peculiar to the United States; others are characteristic of modern medicine generally. We do not want to reject the fruits of progress in scientific understanding, but we sense a peculiar disquietude in living with its increasingly costly, often impersonal, and erratically accessible aspects. When I talk to clinicians over sixty, they are almost unanimously in awe of the extent of change in medicine's technical capacity since they served as residents and house staff. They are well aware of patients treated routinely today whose ailments would have posed intractable challenges forty or fifty years ago. No one wants to throw out the baby of technical capacity—and expectations—with the bathwater of cost, inequity, fragmentation, and bureaucracy.

But what is to be done? Petulance and hand-wringing are not the same as policy. All of us have a stake in the enterprise of medicine, but credentialed professionals have a particular responsibility to act in their own interests and those of the society that has educated them, provided them with effective tools, and granted them respect and a good measure of authority. Perhaps it is time for a fragmented profession to think about what—if anything—binds its members together as a social group and to understand their collective privileges and responsibilities. Such reflections are by no means novel. For centuries physicians have articulated notions of gentlemanly selflessness and service. Codes of medical ethics have ritually invoked such claims to moral stature—as does contemporary bioethics. There has also been a rich body of thought and criticism called social medicine, a tradition of concerned physicians and social commentators who have looked at medicine in an inclusive and integrative way and who have focused on the social ecology of health and disease—on the structural roots of health disparities. It is a way of thinking about medicine in society and society in medicine that dates to at least the end of the eighteenth century and finds its contemporary descendants in social epidemiology and some sectors of the social-science and policy communities.

In the twentieth century, medical sociology and anthropology have also reflected socially critical, often relativist, and implicitly reformist views of health care. Since the 1920s social scientists and patient advocates have warned of the need to understand the individual patient and his or her family—their cultural assumptions and social worlds as well as their manifest symptoms. A generation ago, for example, thoughtful physicians heard much of biopsychosocial approaches, but such socially oriented and integrative approaches have been explicitly or implicitly oppositional since the end of the nineteenth century. The laboratory's cumulative triumphs have made this holistic point of view seem not so much wrong as marginal, elusive, and difficult to study in a systematic way.

Is this dominance of reductionist approaches an inevitable condition of contemporary life? Perhaps this is the reality we face: perhaps this instrumental worldview is as much a part of contemporary medicine and its reward systems as the life-extending and pain-reducing technology that we so much admire. How does one nuance this powerful way of imagining and controlling the world? I hope the following chapters will serve as a cumulative argument for thinking about medicine as a social function, about the ways that healing is not reducible to technical procedures and molecular mechanisms, even if such knowledge remains indispensable. Two terms help me express this point of view: *social efficacy* and *medical citizenship,* and both need explanation.

Social efficacy might seem no more than a well-meaning but meaningless phrase. How can it be evaluated? What is the appropriate metric for the social efficacy of one health-care system as opposed to another? It cannot be measured in terms of indicators such as cost per capita or aggregate morbidity and mortality experience. How do we evaluate the several dimensions of an individual physician-patient interaction? What might its impact be on a family as well as the individual? How serious can I be in invoking something so amorphous as social efficacy—when it is hard enough to evaluate the efficacy of particular drugs or procedures in particular clinical situations? I concede such criticisms but feel that the term has a heuristic—and moral—value. First, it serves to underline the constructedness of all notions of efficacy: the arbitrariness in what we choose to value and to measure and how we choose to measure it—and what we dismiss as unmeasurable and thus—in some sense—valueless; what counts in our social world, in other words, and how we count it. Second, the concept of social efficacy underlines the need to think about the impact of particular practices and policies at a variety of levels and points in time: the individual, the family, the community, the state, and the global community—today and in the future.

A similar set of concerns relates to what I have called "medical citizenship." *Medical* or *biological citizenship* has been used to refer to the negotiated rights of individuals to health and to the integrity of their bodies; but I would like to add another meaning.[7] The term *medical citizenship* can also refer to the peculiar responsibilities, as well as the authority, historically granted the medical profession. That authority is justified in both moral and intellectual terms. Physicians are assumed to act in the best interests of the bodies, families, and dignity of their patients—and thus necessarily the community that has trained them and granted them a good measure of control over the domain of medicine. When I was a child, we learned civics in the hope of making us responsible voters and citizens, and it is particularly incumbent on physicians in this time of instability and change to concern themselves with medicine in its largest social sense—with that part of medicine that cannot be construed in terms of laboratory findings and standard protocols alone. To be a medical citizen is to concern oneself both with the realm of politics and social justice and with clinical judgment. Let me cite Uruguay's president as quoted in the *New York Times* in late 2006. He is a radiologist who has not given up his medical practice: "I don't see working in these two realms as schizophrenic, since both are forms of service to society." "To me politics is an extension of what I do in medicine. . . . But society is also a human organism, and politics is a way of dealing with the pathologies that a society can have. You have to act on that society as you would a human being."[8]

Medicine is not biology, though it applies the laboratory's finding and is legitimated by the power of those findings. Nor is it a market actor alone, though it is certainly that. Medicine has a unique social role and moral history—and identity. Insofar as I have a personal agenda, it is a desire to underline the need to think about medicine in just this way: for physicians to think and act on an understanding of that unique social and moral identity. It means thinking critically about the world beyond technical procedures, the world in which the profession's patients and patients' families live, and the world that informs and constrains clinical choices. At the same time, however, the physician-patient relationship must remain central to the profession's sense of itself—an analytic as well as moral focus of a peculiarly medical literacy, an ability to read society and see how it shapes each interaction between physician and patient. We are in this sense all medical citizens, each one of us capable of using medicine as a way of thinking about society, and of society and politics as ways of understanding medical outcomes.

NOTES

1. Mark Schlesinger, "A Loss of Faith: The Sources of Reduced Political Legitimacy for the American Medical Profession," *Milbank Quarterly* 80 (2002): 185–235.

2. More than a few laypeople believed that yellow fever was spread by some material thing transmitted from person to person—and thus the widespread fear of epidemic disease, even when many learned authorities dismissed the idea of personal contagion.

3. This generalization does not apply as consistently to the patient population of state and private mental institutions founded in the first half of the nineteenth century. Even the wealthy and respectable might consign "insane" and unmanageable family members to hospital care in prebellum America, when they would never have considered doing so for a fever or other visibly somatic ill.

4. See the discussions in "The Tyranny of Diagnosis," chapter 2 in this volume, and in Charles E. Rosenberg, "What Is Disease? In Memory of Owsei Temkin," *Bulletin of the History of Medicine* 77 (2004): 491–505.

5. The preceding New Deal years helped lay the foundations for such activist social policies, but precedents for federal involvement in health and social welfare are much older, ranging from health insurance for seamen in the 1790s to pensions for Civil War veterans and their widows in the late nineteenth century. See Theda Skocpol, *Protecting Soldiers and Mothers: The Political Origins of Social Policy in the United States* (Cambridge, Mass.: Harvard University Press, 1992).

6. Stefan Timmermans and Marc Berg, *The Gold Standard: The Challenge of Evidence-Based Medicine and Standardization in Health Care* (Philadelphia: Temple University Press, 2003); Jeanne Daly, *Evidence-Based Medicine and the Search for a Science of Clinical Care* (Berkeley: University of California Press, 2005); Harry M. Marks, *The Progress of Experiment: Science and Therapeutic Reform in the United States, 1900–1990* (Cambridge: Cambridge University Press, 1997).

7. See, for example, Adriana Petryna, *Life Exposed: Biological Citizens after Chernobyl* (Princeton, N.J.: Princeton University Press, 2002); Keith Wailoo, Julie Livingstone, and Peter Guarnaccia, eds., *A Death Retold: Jesica Santillan, the Bungled Transplant, and Paradoxes of Medical Citizenship* (Chapel Hill: University of North Carolina Press, 2006), esp. pp. 13–14.

8. Larry Rohter, "In Uruguay, the President Also Reads Mammograms," *New York Times*, August 31, 2006, A3.

- -

CHAPTER 2

THE TYRANNY OF DIAGNOSIS

Specific Entities and Individual Experience

DIAGNOSIS HAS ALWAYS PLAYED A PIVOTAL ROLE IN medical practice, but in the past two centuries that role has been reconfigured and has become more central as medicine—like Western society in general—has become increasingly technical, specialized, and bureaucratized. Disease explanations and clinical practices have incorporated, paralleled, and, in some measure, constituted these larger structural changes.

This modern history of diagnosis is inextricably related to disease specificity, to the notion that diseases can and should be thought of as entities existing outside the unique manifestations of illness in particular men and women: during the past century especially, diagnosis, prognosis, and treatment have been linked ever more tightly to specific, agreed-upon disease categories, in both concept and everyday practice. In fact, this chapter might have been entitled "Diagnosis Mediates an Invisible Revolution: The Social and Intellectual Significance of Specific Disease Concepts." It would have been more precise, if somewhat less arresting.

This title also would have the virtue of emphasizing both the importance and comparative novelty of nineteenth- and twentieth-century conceptions of disease, ideas we have come to take so much for granted that they have become invisible. It would not be inappropriate to compare the cultural impact of modern assumptions about the specific nature of disease with the effects of the Newtonian, Darwinian, or Freudian revolutions, "events" that have been long assimilated into the accepted canon of cultural visibility and the subject matter of history textbooks. Certainly this cultural impact is comparable to those conceptual shifts in terms of the ways in which ordinary men and women think about themselves and others. Everywhere we see specific disease concepts being used to manage deviance, rationalize health policies, plan health care, and structure specialty relationships within the medical profession. And I have not even mentioned the countless instances in which clinical interventions and expectations have altered the trajectory of individual lives.

My interest in the history—and historicity—of disease categories began more than a quarter century ago with two incidents fortuitously linked in time. One was my serving as a consultant to a large demographic project studying the causes of death in Philadelphia between 1880 and 1930. The principal investigator faced a methodological dilemma in her critical reading of manuscript death certificates: how were they to code such diagnoses as "old age," "senile," or "marasmus" still common in 1880 but banished by 1930? In previous archival research, I had been struck by early nineteenth-century hospital case records in which either no diagnosis was recorded or general descriptive terms (*fever, fits, dropsy*) served as diagnosis. At about the same time (in the early 1970s) as this coding dilemma, the American Psychiatric Association was in the midst of an embarrassingly public struggle over the revision of its *Diagnostic and Statistical Manual of Mental Disorders*. Most conspicuously, psychiatrists voted, argued, then voted again as they reconsidered the problematic category of homosexuality. Was this a disease or a choice? How could a legitimate disease—in most physicians' minds, a biopathological phenomenon with a characteristic mechanism and a predictable course—be decided by a vote, especially one influenced by feverish lobbying and public demonstrations?[1]

Particularly striking at the end of the twentieth century was the variety of contexts in which we became accustomed to seeing disease concepts being negotiated in public. On September 5, 1997, the *Philadelphia Daily News* reported that a school bus driver in rural Selinsgrove, Pennsylvania, felt, as he put it, like a woman trapped in a man's body and expressed himself by wearing women's clothing, a wig, and eyeliner while driving children to and from school. When anxious parents demanded that he be dismissed, the driver was perplexed: "I don't understand what all the fuss is about. I am diagnosed with gender identity disorder syndrome, and I am being treated."

Gender identity is only one such concept. Contested and widely discussed diagnostic categories, such as attention deficit hyperactivity disorder or chronic fatigue syndrome—not to mention road rage, premenstrual syndrome, and addictions to gambling and sex—have become familiar subjects for public discussion. Physicians are not the only participants in such contentious debates. Not too long ago, to cite another example, feminists challenged the Centers for Disease Control's prevailing definitions of AIDS as ignoring those opportunistic infections peculiar to women. And what are we to make of so-called risk factors such as elevated cholesterol levels or blood pressure or marginal bone densities in postmenopausal women? Are they statistically meaningful predictive indicators, substantive factors in a multicausal etiology, or diseases in themselves?

The social power—and, I should add, utility—inherent in naming diseases is routinely demonstrated in the administrative world of medicine as well as in the wider culture. For example, the Hospital of the University of Pennsylvania, reinvented not long ago as a component of a corporate health-care system, boasted of having created forty evidence-based and cost-containing practice guidelines for what its administrators described as "disease management," comprehensive and regularly updated protocols intended to codify and mandate the practitioner's adherence to formal diagnostic, treatment, prevention, and referral plans. One could cite many parallel instances. During the last two decades of the twentieth century, planners managing America's disjointed health-care system sought to control health-care costs through a variety of bureaucratic schemes organized around such disease-oriented guidelines. Perhaps the most controversial were the diagnosis-related groups that had once been enthusiastically endorsed as a tool for controlling the length of hospital stays. In each of these instances, the presumed existence of ontologically real and definitionally specific disease entities constituted a key organizing principle around which particular clinical decisions could rationally be made. The act of diagnosis-structured practice conferred social approval on particular sickness roles and legitimated bureaucratic relationships. It is not surprising that disease advocacy groups have flourished in the same social and intellectual context, all lobbying for social acceptance and research support. The Alliance of Genetic Support Groups, an umbrella organization, claims a membership of more than 250 such groups organized around genetic diseases alone.

One could provide scores of similar examples, but the moral is apparent. Specific disease categories are omnipresent at the beginning of the twenty-first century, playing substantive roles in a variety of contexts and interactions ranging from the definition and management of deviance to the disciplining of practitioners and the containment of hospital costs. The social uses of disease categories, however, are hardly limited to individual interactions between physicians and their patients or to the setting of research agendas and treatment plans. Philosophers and sociologists of knowledge have voiced an abundance of opinions regarding their epistemological and ontological status, but to the historian, disease entities have become indisputable social actors—real inasmuch as we have believed in them and acted individually and collectively on those beliefs.

Diagnosis is central to the definition and management of the social phenomenon that we call disease. It constitutes an indispensable point of articulation between the general and the particular, between agreed-upon knowledge and its application. It is a ritual that has always linked physician and patient, the emotional and the cognitive, and, in doing so, has legitimated physicians' and the medical system's authority while facilitating particular clinical decisions and providing culturally agreed-upon meanings for individual experience. Not only a ritual, diagnosis is also a mode of communication and thus, necessarily, a mechanism structuring bureaucratic interactions. Although diagnosis has always been important in the history of clinical medicine, it became particularly significant in the late twentieth century with the proliferation of chemical, imaging, and cytological techniques and the parallel conflation of diagnosis, prognosis, and treatment protocols.[2] Diagnosis labels, defines, and predicts, and, in doing so, helps constitute and legitimate the reality that it discerns.

THE SPECIFICITY REVOLUTION

Many of the ways in which we think about disease seem novel, but assumptions about the existence of particular diseases have a long history. Laypersons and physicians have always used words to signify what seemed to constitute discrete disease experiences in their place and time, and such named disease pictures have always been important to practitioners of medicine. In the often-cited language of Knud Faber's classic (1923) history of nosography, the clinician "cannot live, cannot speak or act without the concept of morbid categories."[3] A time- and place-specific repertoire of such agreed-upon disease categories has, in fact, always linked laypersons and medical practitioners and thus has served to legitimate and explain the physician's status and healing practice. Mastering a vocabulary of disease pictures and being able to distinguish among them have long been fundamental to the physician's role, as such knowledge underlies the socially indispensable tasks of diagnosis and prognosis and the rationalization of therapeutic practice.

"Everyone must acknowledge the difficulty of distinguishing diseases," argued the influential Edinburgh teacher and practitioner William Cullen in his widely used late-eighteenth-century textbook of nosology, "but in most cases, the possibility must also be allowed; for whoever denies this, may as well deny that there is such a thing as the medical art."[4] Disease categories have, that is, always linked knowledge and practice, necessary mechanisms for moving between the idiosyncratic and the generalizable, between art and science, between the subjective and the formally objective. And the physician's skills have, as Cullen implied, always turned on differentiating among available clinical pictures. In 1804 Thomas Trotter, a prominent British physician, similarly underlined this ever-present reality. "The name and definition of a disease," he explained, "are perhaps of more importance than is generally thought. They are like a central point to which converging rays tend: they direct future inquirers how to compare facts, and become, as it were, the base on which accumulating knowledge is to be heaped."[5]

Ideas about disease have, in other words, been almost synonymous with the content of medicine as a set of explanatory concepts as well as bedside practices—in Trotter's words, "a base on which accumulating knowledge could be heaped," a mechanism, that is, for converting the uniqueness of experience and particular clinical interactions into a portable and collectively accessible form of data. But not all early-nineteenth-century physicians believed that formal nosologies were worth constructing. "Whether a nosological arrangement, the fruit of modern pathology, is a hopeless expectation, remains yet to be seen," said John Robertson in 1827. "The degree to which diseases are modified by constitution, season, climate, and an infinite variety of accidental circumstances, renders it at least doubtful."[6]

In traditional medicine, disease concepts were focused on the individual sufferer. They were symptom-based, fluid, idiosyncratic, labile, and prognosis-oriented. Diseases were seen as points in time, transient moments during a process that could follow any one of a variety of possible trajectories. A common cold could become bronchitis, for example, and could then resolve without long-term consequences or could terminate, rapidly in a fatal pneumonia or slowly in chronic lung disease. A flux (or looseness of the bowels) could resolve itself without incident or could deteriorate into a fatal or debilitating dysentery. A humoral imbalance might manifest itself in the form of a fever or superficial lesions as the body tried to relieve itself of noxious matter through the skin. The body was always at risk, but a risk configured in idiosyncratic, physiological, multicausal, and contingent terms.[7]

There were a few exceptions. By the beginning of the nineteenth century, epidemic diseases, such as yellow fever, and a number of other ailments, such as smallpox and venereal disease, were often regarded as contagious and thus were thought of somewhat differently. Epidemic outbreaks were explained, however, in terms of either a specific contagion or a peculiar configuration of environmental conditions, with patients' individual constitutions and lifestyles accounting for their differential susceptibility. For example, outbreaks of what in retrospect might be diagnosed as typhus fever were often associated with filthy, crowded, and badly ventilated circumstances, and thus a number of vernacular terms for such epidemics (jail fever, camp fever, famine fever, or ship fever) came into use.

These generally fluid and nonspecific ideas had changed fundamentally by the beginning of the twentieth century. Recognizably modern notions of specific, mechanism-based ailments with characteristic clinical courses were a product of the nineteenth century. Pathological anatomy with its emphasis on localized lesions, physical diagnosis, the beginnings of chemical pathology, and studies of normal and abnormal physiological function all pointed toward the articulation of stable disease entities that could be—and were—imagined outside their embodiment in particular individuals and explained in terms of specific causal mechanisms within the sufferer's body.

Much of this conceptual change had taken place by the 1860s and thus predated the germ theory of disease, which is so often credited with transforming both lay and professional notions of disease as entity. Bright's disease provides an illuminating case in point. It was perhaps the first physician's disease—not only named for a physician in the 1820s but also understood and configured in terms of chemical pathology (albumen appearing in heated urine) and postmortem findings (the visual appearance of abnormal kidneys) as much as with the patient's felt experience and verbal narrative.[8] By the mid-nineteenth century, disease was equated with specificity and specificity with mechanism, all the while decoupling this increasingly ontological conception from idiosyncrasies of place and person. In this sense, the several versions of the idea, articulated in the 1860s and 1870s, that wound infection and communicable disease were caused by living organisms—what has come in retrospect to be called the germ theory—only intensified and documented a way of thinking about disease already widely assimilated, if not consistently applied.

These new ideas became a telling argument for the implicit assumption that disease could be understood as existing in some sense outside the body. Perhaps more fundamentally, germ theories constituted a powerful argument for a reductionist, mechanism-oriented way of thinking about the body and its felt malfunctions. These theories communicated metaphorically the notion of disease entity as ideal type, abstracted from its particular manifestations. A legitimate disease had both a characteristic clinical course and a mechanism, in other words, a natural history that—from both the physician's and the patient's perspective—formed a narrative. The act of diagnosis inevitably placed the patient at a point on the trajectory of that predetermined narrative. Ubiquitous modifying terms such as *atypical* or *complications* only underscored the unspoken centrality of such ideal-typical disease models—and the need for secondary elaborations that would make these concepts more flexible and thus viable clinically.

The second half of the nineteenth century saw the initiation of a trend toward the clinical use of what contemporaries sometimes called "instruments of precision" in the study of disease. One thinks of the thermometer, of blood and urine chemistry and microscopy, and by the 1920s, the blood pressure cuff, the electrocardiogram (EKG) and electroencephalogram (EEG), and x-ray were available in well-equipped hospitals. And all promised to provide ways of describing disease that could be built into tight, seemingly objective pictures, useful in diagnosing and monitoring particular cases yet capable of being generalized into larger understandings. It was not only that temperature could "neither be feigned nor falsified"—as a well-known advocate of the thermometer argued—but also that its results could be expressed in standard units. Thus, the patterned regularity of temperature readings might "aid in the discovery of the laws regulating the course of certain diseases."[9] Similarly, pH readings or red cell counts seemed to provide objective ways of helping characterize an ailment's essential character; aggregated, they promised ever-more-precise understandings of disease as entity.

Disease could now be operationally understood and described. It was measured in units, represented in the visible form of curves or continuous tracings, and taught to successive generations of medical students. Advocates of scientific medicine a century ago did not, of course, think that each of these measures could do more than reflect one characteristic of a particular disease entity. Each individual curve or tracing could in this sense be likened to the particular findings of the blind men who in the well-known fable were asked to describe an elephant. One said it was rather like a snake, another like a tree trunk, and the third—who grasped a tusk rather than the trunk or a leg—likened it to a scimitar. At the beginning of the twenty-first century, some of us might construe this epistemological parable as an argument for the contingent and situated quality of medical knowledge. But to most observers a century ago (and a good many physicians today), the blind men's varied findings were, in sum, proof of the existence of elephants. That is, in a circumstantial way, the gradual deployment of an ever-increasing array of seemingly objective tools worked to establish the texture and corporeality as well as the essential unity of disease entities.

The very possibility of modern epidemiology is in some measure dependent on the acceptance of standardized disease categories as employed in statistics of aggregate morbidity and mortality from hospitals and governments. In the programmatic words of the pioneer vital statistician William Farr in 1837, a uniform nosological nomenclature "is of as much importance in this department of inquiry as weights and measures in the physical sciences."[10] Disease entities were seemingly objective units in which regional difference, social policy, and etiological variables could be weighed. Standard nosological tables seemed to be a necessary tool for helping transcend the subjective, the local, and the idiosyncratic in clinical practice, that is, for implementing what by the late nineteenth century had come to be called the scientific aspect of medicine. Without an agreed-upon vocabulary of disease, for example, the hospital's wards could not contribute to the medical profession's collective task of accumulating valid clinical knowledge.

In addition to their gradual embodiment in the form of accumulated data, accepted disease categories constituted a language that linked physician and patient, especially in the hospital's increasingly bureaucratic context. By the end of the late nineteenth century, the acute-care-oriented hospital had already become a key factor in nurturing an administratively standardized and specific disease-oriented way of thinking about sickness. Disease categories played a fundamental role in the hospital's internal order, and the hospital's increasing centrality served to make diagnosis among a repertoire of specific disease entities indispensable to inpatient medicine and thus the texture of patient experience. That intellectual centrality was intensified and, to a degree, embodied in the beginnings of specialism and the growing significance of the general hospital in delivering care to all classes in society. Moreover, much of the era's systematic clinical investigation was performed in hospitals. Codified in formal classification systems, disease entities became useful tools as these promising institutions sought to at once impose a rational internal order and project an image of efficacy and science. Although in retrospect the late-nineteenth- and early-twentieth-century hospital might seem almost bereft of technology, it had already identified itself as an institution devoted to clinical science, increasingly defined and legitimated by its technological capacities. As I shall argue, those growing capacities were indispensable in providing operational texture to disease as social entity.[11]

In visualizing early-twentieth-century diagnostic practice, one thinks immediately of a hospital-based technology—of machines and microscopes, test tubes and reagents, autoclaves and petri dishes. But nothing illustrates these gradual yet inexorable changes with more circumstantiality than the mundane case record. By the middle of the nineteenth century, the linking of an emphasis on disease nomenclature, hospital practice, and the use of printed forms for recording clinical data was, as an ideal, well established in academic medical circles.[12] By the end of the century, such forms for recording case records had gradually become standard. These documents provided blanks for recording the diagnosis, with little space ordinarily left for a summary of the patient's account of his or her sickness. These uniform case records also included lines for recording the results of blood and urine work and the findings of physical diagnosis. By the 1880s and 1890s, temperature curves were routine components of case records in teaching hospitals, and by the 1920s, tracings from EKGs were often added. Well before the computer emerged to streamline the management of clinical data, the chart at the head of the patient's bed and the hospital's annually aggregated morbidity statistics promised in their different ways to control and rationalize both individual care and the institution's internal order. Diseases were accumulating the flesh of circumstantiality, both biological and bureaucratic. They were becoming social, as well as conceptual, entities.

By the end of the century, the greater coherence and cultural centrality of disease entities were manifested in another, ironic, way. I am referring to their use in the understanding and ordering of behavior. Conspicuous examples of ills such as neurasthenia, hysteria, sexual psychopathy, alcoholism, and homosexuality have already become familiar subject matter for historians. Although they were controversial a century ago—as many of them still are—the cultural work performed by such medicalized categories illustrates the power and pervasiveness of disease entities, no matter how hypothetical, in providing seemingly value-free frameworks for thinking about the normal and the deviant.

As striking as their persistence over time is the way in which such problematic diagnoses were routinely justified in terms of a material mechanism. Without such a mechanism, they could hardly have been advanced as legitimate ills. Here I am referring to the various entities contested during the past century and a half, putative ailments ranging from railroad spine and soldier's heart to shellshock and posttraumatic stress disorder, from neurasthenia to chronic fatigue syndrome. That such diagnoses and their lineal descendants remain contested at the beginning of the twenty-first century is—from the point of view of this chapter—evidence for the persistent cultural centrality of the mechanism-defined disease entity as an explanatory category as much as for the moral and political resonance of these particular would-be ills.

The organization of sickness into discrete categories was consistent as well with the bureaucratic imperative, not only in hospital management, but also in a variety of contexts ranging from life and health insurance to epidemiological and related public health and policy debates. It was not an accident that the 1890s saw agreement on an international classification of causes of death as well as an increasing demand for consistent and comprehensive morbidity statistics.[13]

Disease pictures had already been built into textbooks of medicine; soon they would be sharpened and made even more central under the explicit banner of differential diagnosis. (The origins of the term *differential diagnosis* are obscure. Although it had been used earlier, it is often associated with the didactic efforts of Richard Cabot in the early twentieth century.) The adjective *differential* assumes differentiation among discrete alternatives, and thus it legitimates—and prospectively creates disease entities as social realities, whatever the evidentiary basis for their existence. "By the differential method," Philadelphia teacher John H. Musser wrote unselfconsciously in 1894, "the diagnosis of one of a few possible diseases must be made."[14] Instructing medical students in nosological grammars was an important de facto step in the creation and increasing clinical salience of specific disease entities, because such entities constituted conceptual building blocks around which successive generations of medical students—soon-to-be practitioners—would organize their therapeutic and diagnostic practice. Even earlier in the nineteenth century, the stethoscope and physical diagnosis had promised academic physicians an objective path to understanding the course of particular ills during life. At the beginning of the twentieth century, the clinical-pathological conference provided another institutional and pedagogical ritual. It also underscored the ultimate meaningfulness of discrete disease entities and the social centrality of their diagnosis by focusing on the connection between clinical signs during life and postmortem appearances. The clinical-pathological conference also exemplified and, in part, constituted the dominant role of the hospital in medical education while structuring the relationship between pathology and clinical medicine.[15] In summary, by the end of the nineteenth century, a vocabulary of named disease pictures had already become a widespread and largely unquestioned component of Western medicine.

At the same time, diagnosis of such ills was becoming inexorably and increasingly dependent on tools and techniques derived from the laboratory. This linkage among procedures, machines, and diagnosis seemed to the majority of physicians both desirable and inevitable, for disease could now be defined in increasingly objective terms. It is hardly surprising that as early as the first decade of the twentieth century critics were beginning to express a kind of oppositional disquietude, the fear that a brash and burgeoning scientific medicine meant treating diseases and not people, that it meant excessive dependence on the laboratory's tools and findings, that it meant a glorification of the specialist at the expense of the generalist, and that it denigrated the physician's holistic and intuitive clinical skills.[16]

DISEASE AS SOCIAL ENTITY

All these trends unfolded steadily—if not dramatically—throughout the twentieth century. The narrative that constituted and described each disease became tighter, more procedure-oriented and rule-defined. In the United States, insurance reimbursement reified and intensified this tendency. The logic of clinical epidemiology and randomized clinical trials has also turned historically on the ordering of data in terms of entities, as does much of what has come to be called "evidence-based medicine." The disease entity as concept had, in other words, steadily accumulated the texture of bureaucratic and biological circumstantiality.

The key factors are obvious. One is technology, medicine's increasing ability to interrogate and even alter the trajectory of particular disease pictures. In the twentieth century, both therapeutic innovation and a growing diagnostic capacity have defined and legitimated disease concepts as they have empowered medical practitioners and reconfigured lay expectations of medicine. Such innovations have even altered the ecology and manifestations of disease: after antibiotics, bacterial pneumonias were in some real sense not the same entities; and diabetes after insulin therapy, pernicious anemia after liver extracts, chronic kidney disease after dialysis, and heart disease after angioplasty, all became new diseases, given shape, texture, and often a greater degree of predictability through the agency of medicine even when they could not be definitively cured.

Particular ills have often been given a definitional specificity through the specificity of their response to therapeutics. Pernicious anemia, for example, was defined in the 1920s as the part of the spectrum of anemias that responded to liver extracts.[17] The role of lithium in helping define and legitimate bipolar disorder provides a familiar parallel example, as does the role of quinine in differentiating malaria from other recurrent fevers. It should be remembered that the predictability of a response to a particular agent implies the specificity of the pathological mechanism and hence its epistemological legitimacy. This circular—and self-fulfilling—tightness of fit has historically provided evidence for the hard, sharply bounded, and mechanism-legitimated definition of disease entities. It is instructive, as well as ironic, that contemporary forms for ordering clinical tests often specify a presumed diagnosis, prospectively justifying the laboratory expenditure.

The increasing dominance of the twentieth-century hospital as a site for research, education, and the delivery of care was a second key factor in the social embodiment of disease. By the end of the nineteenth century, the hospital had emerged as a dynamic site for the delivery of urban health care and for the development of elite medical careers.[18] The growth of clinical pathology, imaging, and other diagnostic tools have not only helped centralize care in the hospital but have also helped operationalize and embody disease entities. Late-twentieth-century imaging, immunological, and cytological procedures provide even more precise assurances that clinical medicine can base diagnoses on an understanding of the body's fundamental mechanisms and not simply the externally observable or patient-reported signs and symptoms of disease. Disease entities became more plausible, more sharply defined, and more frequently the framework and rationale for predetermined therapeutic interventions. Once articulated, these entities have helped order the relationships among machines, experts, caregivers, and patients in the hospital, creating a structure of seemingly objective priorities and practices. They have provided a language as well, enabling and structuring communication among different sectors of the health-care system: to what service would a patient be assigned; what sequence of tests or procedures was most appropriate; and, in America at least, what procedures would be reimbursed.

Bureaucratic structures and practices constituted a third key aspect of the twentieth-century embodiment of disease entities. Bureaucracy has, moreover, become increasingly dependent on the deployment of numbers and categories; thus data management provides another kind of tissue in a late-twentieth-century disease's social body. Advocates of the computer in the clinic have worked eagerly to digitize, rationalize, and ultimately help link diagnosis, prognosis, and therapeutics, intensifying a tendency already well under way before the computer era.[19] Together, randomized clinical trials, consensus conferences, and the coding conventions of nosological tables—as exemplified in the *Diagnostic and Statistical Manual* of the American Psychiatric Association—create socially agreed-upon parameters of disease.[20] So, too, does the way in which laboratory findings are often expressed in numbers, stages, and thresholds. The bureaucratic need for numbers that legitimate and trigger a sequence of additional diagnostic, therapeutic, and administrative actions also obscures the very constructedness of those numbers. The fact that such numbers are routinely generated by seemingly objective, highly technical tools and procedures works to endorse their plausibility and meaningfulness. Ironically, the very negotiated quality of the numerical values that define ailments and specify their treatments creates a reciprocal social rigidity as numbers become the measure and legitimation of presumed things. Participants in the health-care system are well aware of the pitfalls (not the least of which is a loss of autonomy) inherent in the use of such operationalized definitions, but they nevertheless remain in thrall to the need for seemingly objective measures through which to manage disease both therapeutically and administratively.

The use of ideal-typical disease pictures creates experience as well as conceptualizes and records it. The power of specific disease entities rests not in their Platonic—abstract—quality but in their ability to acquire social texture and circumstantiality, structure and legitimate practice patterns, shape institutional decisions, and determine the treatment of particular patients. We see disease entities given social life in the use of treatment protocols in research and cure; we see it in the use of what has come to be called "evidence-based medicine"; we see it embedded in expert software and elaborate treatment guidelines. Protocols are powerfully constraining even as physicians concede their frequent arbitrariness in particular clinical situations.

Medical knowledge is consistently articulated around disease pictures. They not only help make experience machine readable but also help create that experience. Disease categories connect aggregate statistical data and practice. As we have seen, they link and conflate diagnosis, prognosis, and treatment; they are ghosts in the health system's software. But perhaps ghosts are an imprecise metaphor, for systems of disease classification are very real and quite intractable technologies, linguistic tools that allow the machines and institutions of government and health care to function. Disease entities are social realities, actors in complex and multidimensional negotiations that configure and reconfigure the lives of real men and women. Just as disease can be created by ideological and cultural constraints in traditional societies—as generations of anthropologists have reminded us—so contemporary medicine and bureaucracy have constructed disease entities as socially real actors through laboratory tests, pathology-defining thresholds, statistically derived risk factors, and other artifacts of a seemingly value-free biomedical scientific enterprise.

PARADOXES OF DISEASE SPECIFICITY

This way of thinking about disease—the vision of abstracted disease entities as ever-more-precise mirrors of nature—has become extraordinarily pervasive, yet in its very explanatory power it has posed a variety of intractable social dilemmas, problems that in fact underline the cultural centrality and ubiquitousness of contemporary disease concepts.[21]

One such problem is implicit in the way in which we use disease categories to perform the cultural work of enforcing norms and defining deviance. A second dilemma grows out of the difficulty inherent in fitting idiosyncratic human beings into constructed and constricting ideal-typical patterns, patterns necessarily abstract yet, in individual terms, paradoxically concrete. A third problem lies in medicine's growing capacity to create proto-disease and disease states that shape everyday medical practice and thus individual lives. Fourth is what might be described as the bureaucratic imperative, the way in which the creation of nosological tables, guidelines, protocols, and other seemingly objective and practice-defining administrative mechanisms constitutes in aggregate an infrastructure mediating between and among government and the private sector, practitioners and patients, specialists and generalists, and—in the United States—insurers and providers. That infrastructure is as much a part of the experience of sickness as diagnosis or clinical management; they are in fact indistinguishable.

Since the mid-nineteenth century, putative disease entities have been called on to do a variety of cultural tasks, most conspicuously to naturalize and legitimate conceptions of difference and deviance. I am referring, of course, to an assortment of problematic ailments ranging from attention deficit disorder to homosexuality to alcoholism. Not surprisingly, such entities remain controversial because their diagnosis turns on underlying conceptions of normal behavior as well as individual responsibility and professional jurisdiction. We have grown accustomed to the public and often contentious negotiations surrounding these and other problematic disease categories—with individuals, advocacy groups, and medical specialty associations all participating. Equally predictable is the way in which successive generations of physicians have advanced somatic mechanisms to legitimate and explain such ills.

The history of forensic psychiatry in the past century and a half reflects, for example, successive iterations of the notion that free agency could be inhibited by some biopathological process—such as moral insanity in the mid-nineteenth century and a variety of successor diagnoses—that overrode an offender's ability to have chosen the right and rejected the wrong. Insofar as the supposedly pathological behaviors can be construed as the consequence of a somatically based—and thus deterministic—mechanism, such entities necessarily undermine traditional notions of agency and so engender both legal and ideological conflict.

I have described a recurring paradox: the unavoidable use of reductionist means to achieve cultural and behavioral—necessarily holistic, multidimensional, and contingent—ends. Sociologists have described one aspect of this history as the medicalization of deviance. One can hardly disagree, but what they have generally failed to emphasize is this very paradox: the consistent use of determinist, mechanism-oriented explanatory strategies to define, stigmatize, and destigmatize. In this sense, one can trace a direct line of intellectual descent from such nineteenth-century formulations as George Beard's neurasthenia (a constitutional nervous weakness that could manifest itself in a variety of ways) or Cesare Lombroso's degeneracy (an evolutionary atavism) to late-twentieth-century notions of the genetic determinism of criminality, homosexuality, or even dyslexia, depression, and risk taking.[22] The ways in which such speculative entities serve as vehicles for articulating cultural norms have, in fact, become a cliché in contemporary historical writing. But the persistent framing and reframing of such entities in terms of somatic mechanisms are as striking as the failure of such formulations to compel general and lasting assent. What has remained consistent in the last century and a half is the form of such norm-defining and -enforcing sanctions: the creation of hypothetical, mechanism-based disease entities and the role of credentialed experts in attesting to the validity or illegitimacy of such ills. Even when we cannot agree on an underlying biological mechanism, we find ourselves debating the existence and legitimacy of scores of problematic ailments. For many participants, the diagnosis constitutes a kind of social equity, although others may regard the same designation as a form of stigmatization. Contemporary discussions of chronic fatigue syndrome and "chronic" Lyme disease constitute particularly apt examples.[23]

A somewhat different set of dilemmas turns on the difficulty inherent in adjusting the individual to the general and the abstract. How is a particular case of tuberculosis or lupus, for example, related to the textbook's description or a treatment protocol's prescriptions? Agreed-upon disease pictures are configured in contemporary medicine around aggregated clinical findings—readings, values, thresholds—whereas therapeutic practice is increasingly and similarly dependent on tests of statistical significance. Yet men and women come in an infinite variety, a spectrum rather than a set of discrete points along that spectrum. An instance of cancer, for example, exists along such a continuous spectrum; the staging that describes and prescribes treatment protocols is no more than a convenience, if perhaps an indispensable one. In this sense, the clinician can be seen as a kind of interface manager, shaping the intersection between the individual patient and a collectively and cumulatively agreed-upon picture of a particular disease and its optimal treatment.

Within this managerial context, the practitioner's role is inevitably compromised and ambiguous. On the one hand, the physician's status is enhanced by providing access to the knowledge and techniques organized around disease categories. At the same time, however, the physician is necessarily constrained by the very circumstantiality of that generalized knowledge, by the increasing tightness of diagnostic and treatment guidelines (and, in the United States in recent years, by the mixture of malpractice angst and managed care that impinge ever more powerfully on autonomy in clinical choice). Although this pattern of practice is described and justified as "ensuring quality," in the terminology of contemporary health administration, slippage, frustration, communication failure, and unmet expectations are inevitable. How, for example, does one explain a prognosis framed in aggregate probabilities to a particular patient and his or her family, when every patient constitutes an N of 1? How does one ensure clinical flexibility and an appropriate measure of practitioner autonomy in such a system? How does one manage death—which is not precisely a disease—when demands for technological ingenuity and activism are almost synonymous with public expectations of a scientific medicine?

Finally, there are the much-discussed moral and policy implications of an acute care– and mechanism-oriented clinical medicine that assigns a comparatively low priority to the multicausal, to the social, ecological, public-policy, and quality-of-life perspectives. Western medicine's historical focus on specific disease entities and the management of acute illness is obviously an integral aspect and product of this fundamental worldview, and thus policy. This is another area of maladjustment or difficulty of fit, not, from this perspective, the fit between the individual patient and the generalized disease picture, but between a reductionist, mechanism-centered understanding of disease and a collective strategy for defining and maximizing health. This pattern of episodic care structured around specific entities seems particularly problematic, moreover, in an age of chronic illness—when men and women do in fact die of old age.

Another dilemma growing out of the emphasis on disease specificity turns on our increasing ability to create and modify disease entities, what one might call the iatrogenesis of nosology. One significant aspect of this technology-dependent process is the invention of protodiseases, for example, elevated blood pressure or cholesterol levels or low bone densities in postmenopausal women. Once articulated and disseminated in practice and the culture generally, these conditions become emotional and clinical realities, occupying a position somewhere between warning signal and pathology. Our expanding armamentarium of cytological, biochemical, and physiological function, and imaging tests create screening and treatment options and thus new and altered diseases. Prostate cancer, for example, has been changed as a social and clinical reality by the availability of new screening options, as has breast cancer by mammography. Genetic testing has already created new diseases-carrier states in Huntington's chorea or Tay-Sachs, for example, and promises to shape a multiplicity of such immanent ailments. Breast cancer figures prominently in this prognostic context as well.[24]

The bureaucratic management of disease creates another kind of dilemma. Nosological categories play an indispensable administrative role. In one of its aspects, disease *is* its bureaucratic management. Can one imagine today's medicine and society without a rationalizing and organizing vocabulary of disease entities? Such nosologies constitute a tool for arranging compromises among a variety of interested actors—let us say patients, drug companies, competing specialties, insurers, researchers, and patients—as well as a tool for the day-to-day administration of such compromises. In this functional sense, modern society might be thought of as demanding the creation of diseases as social entities; they help legitimate the society's values and status arrangements and also provide an instrument for the system's day-to-day management.

In this sense, a repertoire of disease categories becomes a mediating interface between and among parts of the system. Once articulated, such bureaucratic categories cannot help but exert a variety of substantive effects on individuals and institutional relationships. Because most somatic disease categories seem in themselves value-neutral, for example, and thus legitimate care, there seems something wrong in not treating the sick when an efficacious technology is available and mandated by a particular diagnosis. Thus, a poor or homeless person becomes visible to the health-care system when diagnosed with an acute ailment but then returns to invisibility once that episode has been managed. It is almost as though the disease, not its victim, justifies treatment. The employment of seemingly objective disease categories thus obscures the conflicted relationships among medicine's moral, technical, and market identities.

Linkage is the key concept here, the way in which bureaucracy, market, cultural identity, and other factors all can interact around the creation of an agreed-upon disease threshold. One can cite scores of instances illustrating this proliferating phenomenon. In 1999, for example, the National Institutes of Health broadened its definition of overweight, conceptualizing their new categories in terms of a body mass index. According to a *New York Times* story published on May 2, "Until last year, men were overweight if their body mass index was 27.3 or higher; for women, the cut-off was 27.8. Today anyone with a body mass index of 25 or over is overweight by government standards. And a new category, obese, has been added for those whose index is 30 or above." With, as the *Times* story put it, "excess body fat increasingly being viewed as a disease," the drug industry could now market a "pharmacological fix." Pharmaceutical firms could now hope for an expanded market for diet drugs, legitimated for insurance reimbursement by their status as "disease treatment." Somewhat more recently, to cite a parallel example, the Food and Drug Administration announced that it had approved an already much-prescribed antidepressant, Zoloft, for use in treating posttraumatic stress syndrome in women.[25] The drug could now be advertised for this use, and, presumably, its costs covered by insurers.

THE SOCIAL FUNCTION OF DIAGNOSIS

All these problems illustrate the central role of diagnosis itself. Perhaps most fundamentally, the act of diagnosis links the individual to the social system; it is necessarily a spectacle as well as a bureaucratic event. Diagnosis remains a ritual of disclosure: a curtain is pulled aside, and uncertainty is replaced—for better or worse—by a structured narrative. Think of the moment when the patient and his or her family are shown revelatory x-rays or printouts, when the physician pronounces a diagnosis that can allay or intensify fear. Even though contemporary diagnosis is ordinarily a collective, cumulative, and contingent process, it is significant that most of us think of it as a discrete act taking place at a particular moment in time. Both physician and patient are hostage to this age-old ritual. There is an instructive irony in the way in which nosological tables can effectively reshape the lives of particular men and women, even when the physician bestowing a particular diagnosis is aware of how arbitrary that determination may be.

Disease pictures are formally objective narratives that both provide meaning and underline social hierarchies. We ordinarily expect a diagnostic determination and subsequent communication as part of the clinical interaction; even its absence shapes our expectations and becomes part of a necessarily altered narrative. Such diagnoses need not be optimistic to be socially efficacious. In our largely secular society, sometimes meaning can be reduced to the admission that we do not yet know the mechanism, that medicine cannot intervene in certain predetermined illness trajectories, that some disorders remain mortal. Anxiety and mystery can be ordered, if not entirely allayed

Diagnosis remains both a bureaucratic and an emotional necessity—for records, for reimbursements, and for the coordination of complex intraprofessional and institutional relationships. Vocabularies of disease entities are a necessary aspect of the contemporary world and are not easily disaggregated from the technical capabilities that the great majority of us have come to expect from medicine. When America's political and medical spokespersons boast of enjoying the world's best health care, they are implicitly referring to the capacity to intervene in the trajectory of disease, to alter a worst-case scenario.

Our understanding of the biopathological aspects of disease and the technologies available to manage and understand them are a part of reality as much as are our clogged arteries or dysfunctional kidneys. It is in this sense that I employ the phrase *tyranny of diagnosis*. I might just as well have used the term *indispensability*. Diagnosis is a cognitively and emotionally necessary ritual connecting medical ideas and personnel to the men and women who are its clients. Such linkages between the collective and the uniquely individual are necessary in every society, and in ours the role of medicine is central to such negotiated perceptions and identities. The system of disease categories and diagnosis is both a metaphor for our society and a microcosm of it. Diagnosis is a substantive element in this system, a key to the repertoire of passwords that provide access to the institutional software that manages contemporary medicine. It helps make experience machine readable.

In the act of diagnosis, the patient is necessarily objectified and recreated into a structure of linked pathological concepts and institutionalized social power. Once diagnosed, that bureaucratic and technically alienated disease-defined self now exists in bureaucratic space, a *simulacrum* thriving in a nurturing environment of aggregated data, software, bureaucratic procedures, and seemingly objective treatment plans. The power of the bureaucratized diagnostic function is, as I have suggested, exemplified in the willingness of physicians to employ the constraining—yet empowering—categories of such nosologies even when they remain skeptical of their validity. The routine use by clinicians of the American Psychiatric Association's *Diagnostic and Statistical Manual* and its often arbitrary categories remains a powerful example of this phenomenon. Equally revealing is the spectacle of individual sufferers and disease interest groups demanding the attribution of particular disease identities, of which chronic fatigue syndrome is a particularly visible example.

Elements of this argument have become familiar in the past generation. They are manifestations of a more general antireductionist critique widely articulated and elaborated since the beginning of the twentieth century. Such criticisms have in fact become a cliché. A wary skepticism of the laboratory, of the impersonal acute-care hospital, and of a dehumanizing specialism has had a history as long as the twentieth century itself. It has in fact become fashionable among humanistic and social science–oriented commentators to dwell on the distinction between illness and disease, between the patient's felt experience and the constructions placed on that experience by the world of medicine.[26] This distinction is certainly valuable for the purposes of analysis, but in practice sickness is, of course, a mutually constitutive and interactive merging of the two; we are not simply victimized, alienated, and objectified in the act of diagnosis. Disease categories provide both meaning and a tool for managing the elusive relationships that link the individual and the collective, for assimilating the incoherence and arbitrariness of human experience to the larger system of institutions, relationships, and meanings in which we all exist as social beings.

Thus, specific disease entities can be understood as holistic and integrative in a social-system sense, just as they can be fragmenting and alienating in terms of an individual's relationship to that larger society. We are never illness or disease but, rather, always their sum in the world of day-to-day experience. Illness and disease are not closed systems but mutually constitutive and continuously interacting worlds. In the patient's case, there is always experience as well; we are always in contact with our own worlds of physical and emotional pain and experience—and thus identity—that cannot be reduced to the external zone of intersection between society and the men and women who constitute it. Identity relates to individual consciousness as well as social location. Pain, sickness, and death help make that particular aspect of experienced identity unavoidable and, at some level, ultimately inaccessible to medicine's changing understandings of disease and the tools for managing it.

A preliminary version of this chapter was originally presented as the Fifth Amalie and Edward Kass Lecture at the Wellcome Institute for the History of Medicine, London, June 22, 1999. I appreciate the opportunity offered by the Wellcome Institute and the many helpful comments I received from listeners and subsequent readers. I would particularly like to thank Robert Aronowitz, Chris Lawrence, and Steven Peitzman for helpful suggestions that have been incorporated into the text.

1. See Ronald Bayer, *Homosexuality and American Psychiatry: The Politics of Diagnosis* (New York: Basic Books, 1981); Gerald Grob, "Origins of DSM-1: A Study in Appearance and Reality," *American Journal of Psychiatry* 148 (1991): 421–31; Stuart A. Kirk and Herb Kutchins, *The Selling of DSM: The Rhetoric of Science in Psychiatry* (New York: Aldine de Gruyter, 1992); Herb Kutchins and Stuart A. Kirk, *Making Us Crazy: DSM: The Psychiatric Bible and the Creation of Mental Disorders* (New York: Free Press, 1997).

2. See Nicholas A. Christakis, "The Ellipsis of Prognosis in Modern Medical Thought," *Social Science and Medicine* 44 (1997): 301–15, and *Death Foretold: Prophecy and Prognosis in Medical Care* (Chicago: University of Chicago Press, 1999).

3. Knud Faber, *Nosography in Modern Internal Medicine* (New York: Paul B. Hoeber, 1923), vii.

4. William Cullen, *Nosology; Or, a Systematic Arrangement of Diseases* (Edinburgh: William Creech, 1800), ix.

5. Thomas Trotter, *Medicina Nautica: An Essay on the Diseases of Seamen,* 3 vols., 2nd ed. (London: Longman, Hurst, Rees, and Orme, 1804), 3:467.

6. John Robertson, *Observations on the Mortality and Physical Management of Children* (London: Longman, Rees, Orme, Brown and Green, 1827), 82–83.

7. Charles E. Rosenberg, "The Therapeutic Revolution: Medicine, Meaning, and Social Change in Nineteenth-Century America," *Perspectives in Biology and Medicine* 20 (1977): 485–506.

8. The term *Bright's disease* is now obsolete and does not correspond to today's diagnostic terminology. It was, however, used as recently as the mid-twentieth century and reflects a contemporary understanding of renal pathology. See Steven J. Peitzman, "Bright's Disease and Bright's Generation: Toward Exact Medicine at Guy's Hospital," *Bulletin of the History of Medicine* 55 (1981): 307–21 and "From Bright's Disease to End-Stage Renal Disease," in *Framing Disease: Studies in Cultural History,* ed. Charles E. Rosenberg and Janet Golden (New Brunswick, N.J.: Rutgers University Press, 1992).

9. C. A. Wunderlich and Edward Seguin, *Medical Thermometry, and Human Temperature* (New York: William Wood, 1871), vii.

10. William Farr, *Registration of the Causes of Death: Regulations and a Statistical Nosology; Comprising the Causes of Death, Classified and Alphabetically Arranged* (London: W. Clowes for Her Majesty's Stationery Office, 1843), app. p. 6; see also John

M. Eyler, *Victorian Social Medicine: The Ideas and Methods of William Farr* (Baltimore: Johns Hopkins University Press, 1979); and Theodore M. Porter, *Trust in Numbers: The Pursuit of Objectivity in Science and Public Life* (Princeton, N.J.: Princeton University Press, 1995).

11. Charles E. Rosenberg, *The Care of Strangers: The Rise of America's Hospital System* (New York: Basic Books, 1987).

12. See Robert D. Lyons, *A Handbook of Hospital Practice; Or, an Introduction to the Practical Study of Medicine at the Bedside* (New York: Samuel S. and William Wood, 1861).

13. See "Contested Boundaries," chapter 3 in this volume.

14. John Herr Musser, *A Practical Treatise on Medical Diagnosis for Students and Physicians* (Philadelphia: Lea Bros., 1894), 19.

15. Russell Charles Maulitz, "Pathology," in *The Education of American Physicians: Historical Essays,* ed. Ronald L. Numbers (Berkeley: University of California Press, 1980), 122–42.

16. See Christopher Lawrence, "Incommunicable Knowledge: Science, Technology and the Clinical Art in Britain, 1859–1914," *Journal of Contemporary History* 20 (1985): 503–20; and Christopher Lawrence and George Weisz, eds., *Greater than the Parts: Holism in Biomedicine, 1920–1950* (New York: Oxford University Press, 1998).

17. See Keith Wailoo, *Drawing Blood: Technology and Disease Identity in Twentieth-Century America* (Baltimore: Johns Hopkins University Press, 1997).

18. Rosenberg, *Care of Strangers;* Rosemary Stevens, *In Sickness and in Wealth: American Hospitals in the Twentieth Century* (New York: Basic Books, 1989); see also Steve Sturdy and Roger Cooter, "Science, Scientific Management and the Transformation of Medicine in Britain c. 1870–1950," *History of Science* 36 (1998): 421–66.

19. See Marc Berg, *Rationalizing Medical Work: Decision-Support Techniques and Medical Practices* (Cambridge, Mass.: MIT Press, 1997).

20. See Harry M. Marks, *The Progress of Experiment: Science and Therapeutic Reform in the United States* (Cambridge: Cambridge University Press, 1997); and J. Rosser Matthews, *Quantification and the Quest for Medical Certainty* (Princeton, N.J.: Princeton University Press, 1995).

21. It is not my intent to question or defend the pragmatic usefulness or the epistemological status of disease entities by placing them in this contingent historical framework. Rather, I am trying to address a different problem: understanding that framework. Perhaps most important, I do not want to impugn their provisional value in increasing the understanding of the body in health and disease. Efforts to contextualize the enterprises of science and medicine are often construed as relativist and delegitimating, but I would suggest that such defensive reactions are created by polarized ideological positions, not by the demands of logic. To historicize and contextualize our changing understandings of disease is not the same as impugning their ontological basis—or implying that all contingent positions occupy a symmetrically arbitrary relationship to the natural world. I have

made related arguments elsewhere: "Disease and Social Order in America: Perceptions and Expectations" and "Framing Disease: Illness, Society, and History," both in Charles E. Rosenberg, *Explaining Epidemics and Other Studies in the History of Medicine* (Cambridge: Cambridge University Press, 1992).

22. Daniel Pick, *Faces of Degeneration: A European Disorder c. 1848–c. 1918* (Cambridge: Cambridge University Press, 1989).

23. Robert Aronowitz, *Making Sense of Illness: Science, Society, and Disease* (Cambridge: Cambridge University Press, 1998); see also Hillary Johnson, *Osler's Web: Inside the Labyrinth of the Chronic Fatigue Syndrome Epidemic* (New York: Crown Books, 1996).

24. The contemporary debates over the efficacy of screening mammography for breast cancer, for example, underscore the complexity and multidimensionality of such issues. Social expectations and economic imperatives as well as rapidly changing technology and our still imperfect understanding of the ailment's natural history all interact to configure a particular, time-specific screening reality. Changes in any of these component elements imply change in the aggregate.

25. *Philadelphia Inquirer,* December 6, 1999.

26. Arthur Kleinman, *The Illness Narratives: Suffering, Healing, and the Human Condition* (New York: Basic Books, 1988).

- -

CHAPTER 3

CONTESTED BOUNDARIES

Psychiatry, Disease, and Diagnosis

SOME YEARS AGO THE *NEW YORK TIMES* FRONT PAGE reported the outcome of a much-discussed courtroom drama, the Andrew Goldstein murder trial. Previously diagnosed and treated—or more often not treated—as a chronic schizophrenic, Goldstein had killed a randomly chosen young woman by pushing her in front of a subway train. Despite his unchallenged diagnosis, the jury convicted Goldstein of second-degree murder. "He seemed to know what he was doing," one juror said after the trial. "He picked her up and threw her. That was not a psychotic jerk, an involuntary movement." Another juror explained that they had thought the defendant was "in control and acted with intent to kill." "It was staged and executed," he said of the lethal attack. "There was forethought and exquisite timing."[1]

This is a story of intellectual and institutional conflict, of inconsistent conceptions of disease and impulse control, and of a chronically ill-starred relationship between law and medicine. It is also a story that might have been written in 1901 as well as in 2001, and, of course, the formal categories of the cognitively defined right-and-wrong test for criminal responsibility still lingers in most American courtrooms. Not too long ago, our media retailed the story of Houston mother Andrea Yates, who drowned her five children. The local prosecutor argued that there was no question concerning her responsibility. She knew right from wrong, he said. "You will also hear evidence that she knew it was an illegal thing, that it was a sin, that it was wrong."[2]

Such highly publicized forensic dramas represent just one—in some ways far from typical—example of a much larger and more pervasive phenomenon: the negotiation of disease in public and the particularly ambiguous status of hypothetical ailments whose presenting symptoms are behavioral or emotional. Most of us would agree that there is some somatic mechanism or mechanisms (whatever their nature or origin) associated with grave and incapacitating psychoses, but, as the dilemma of criminal responsibility illustrates, even in such cases we remain far from agreeing about management and precise disease boundaries. There is, however, a much larger group of individuals who represent a more elusive and ambiguous picture. They are men and women who experience incapacitating emotional pain, who have difficulties in impulse control, or who—even if they have not violated a criminal statute—behave in ways that seem socially or morally unacceptable to many of the people with whom they come into contact.

Sociologists and social critics have, for more than a quarter century, spoken of the medicalization of deviance, of the tendency to recategorize sin(s) as pathology(ies) and consign the management of such conditions to appropriately certified practitioners.[3] But this is only one subset of a larger phenomenon, one that is in a literal sense coterminous with the history of medicine as a specialized calling. I refer to the assignment of certain aspects of human pain and incapacity to the realm of medicine and to the physician's care and explanatory authority. "Medicalization" might perhaps be better understood as a long-term trend in Western society toward reductionist, somatic, and—increasingly—disease-specific explanations of human feelings and behavior as well as unambiguously physical ills.

Nevertheless, the phenomenon remains complex, inconsistent, and contingent—even if expansive and increasingly pervasive. The relationships among disease concepts and painful or socially problematic behaviors have been and are being contested and recontested—not only in melodramatic courtroom situations, but in countless clinical, bureaucratic, and administrative contexts. Moreover, deviance is hardly a discrete and objective thing; it is time, place, and even class specific. Think of now casually accepted sexual behaviors, ranging from masturbation to "excessive" female sexuality, that a century ago would have been seen as certainly deviant and possibly pathological. A bright line between disease and willed misbehavior or culpable self-indulgence or idiosyncratic emotional discomfort will not easily be agreed upon, while the cultural and bureaucratic need to create such boundaries will hardly disappear. Meanwhile individual men and women, lay and professional, act out complex and not always consistent agendas shaped by personal, familial, generational, and social locational realities.

Criminal responsibility and the vexed relationship between law and medicine constitutes only one such area of recurring negotiation. The media provide countless instances of such public controversy, only a very small minority of which are acted out in criminal courts. I need only refer to a number of problematic categories, entities such as gender identity disorder, attention deficit hyperactivity disorder, social anxiety disorder, or chronic fatigue syndrome, not to mention road rage and premenstrual syndrome or putatively—and thus potentially exculpatory—pathological addictions to gambling and sex. The outpatient psychiatric facility at Maclean Hospital in Belmont, Massachusetts, recently offered treatment for "computer addiction"—an inability to refrain from the Internet. Readers of their brochure were informed that the hospital offered a full range of "specialized clinical programs in addition to those for computer addiction . . . [including] . . . those dealing with anxiety, depression, alcohol and drug abuse, Alzheimer's, dementia, personality disorders, bipolar and psychotic illnesses, dissociative disorders, trauma, sleep disorders, human sexuality, and women's and men's issues."

Public policy in regard to drug and alcohol use represents another tenaciously contested field for debating the applicability and legitimacy of disease concepts; billions of dollars and many thousands of lives have been altered by deeply felt and widely disseminated assumptions concerning what has come to be called substance abuse. Are such behaviors the symptoms of a chronic disease (with a biochemical and perhaps genetic substrate) that demands treatment? Or are they crimes to be punished? "Addiction is a chronic disease that demands a medical and public health response," one advocate of the disease model contended in a typical letter to the editor: "It is not a moral lapse."[4] Such conflicts surrounding the ontological status—and thus social legitimacy—of behavioral and emotional ills have been endemic since their widespread articulation well over a century ago. Patterns of cultural and clinical visibility change, but that ambiguity remains. Today, for example, problems of mood constitute a particularly pervasive and diagnostically tendentious category: is depression a thing, a dimension of human diversity and the human condition, an appropriate response to situational realities, or some combination of them? These questions have become all too familiar in the past generation.

Perhaps the most embarrassingly public of such debates over the epistemological legitimacy of a disease category took place more than a generation ago and was occasioned by a planned revision of the American Psychiatric Association's *Diagnostic and Statistical Manual (DSM)*. In retrospect, the most egregious aspect of this conflict was the series of votes surrounding a reconsideration of the problematic category "homosexuality." Was it a disease or a choice? And how could a legitimate disease—in most physicians' minds a biological phenomenon with a characteristic mechanism and predictable course—be decided by a vote—a vote, moreover, that was influenced by feverish lobbying and public demonstrations?[5] Although it has become routine, many Americans still find it unseemly that diagnoses should be shaped in part by advocacy groups and websites, that disease-targeted research funding should be determined in part by lobbyists, lay advocates, and journalists and not the seemingly objective and inexorable logic of laboratory findings. And if lobbying for federal support of cancer research and treatment has come to seem no more than the normal way we negotiate public policy, a similar worldly judgment seems not so natural when applied to ailments of mood and behavior. In the private sector during the past half century, we have seen how pharmaceutical industry research and marketing decisions have helped reshape both medical and lay notions of emotional illness and its treatment, but we have also witnessed the articulation of a vigorous critique of such trends, a criticism of not only specific corporate tactics but also the social role of business, its relationship to government, and the problematic nature of psychiatry's diagnostic categories.[6]

In this chapter I will outline the key characteristics of an era of expanding nosological boundaries—beginning in roughly the last third of the nineteenth century and extending into the present—and then specify some of the reasons that controversy often continues to surround disease categories that promise to explain behavior and emotional pain.[7] Much of what I will discuss falls into psychiatry's domain of clinical responsibility. The psychiatrist has been for more than a century the designated trustee of those social and emotional dilemmas that can plausibly—and thus usefully—be framed as the product of disease. We contest the precise definitions and appropriate clinical and social responses to somatic ills as well, but behavioral and emotional ailments constitute a particularly sensitive and contingent subset of problems. Psychiatry has, since its origins as a specialty in the nineteenth century, been a would-be definer of boundaries, a delineator and designated manager of the normal and abnormal and thus unavoidably a key participant in this never-ending debate. At the same time it has suffered from a recurrent status anxiety—one might call it procedure envy or organic inferiority. Psychiatry has been chronically sensitive to its inability to call upon a repertoire of tightly bounded—seemingly objective and generally agreed-upon—diagnostic categories based firmly on biopathological mechanisms.[8] Ailments such as pellagra and paresis that had been the psychiatrist's responsibility when their cause and treatment were obscure, left the specialty's domain when their mechanisms were clarified and effective treatment established. Psychiatry remains the legatee of the emotional, the behavioral, and the imperfectly understood. In this sense it has been a poor relation of its specialist peers in surgery and internal medicine.[9] And one need hardly speak of psychiatry's history of uncritical flirtations with seemingly effective somatic interventions. Think, for example, of insulin shock and lobotomy.

THE SPECIFICITY TRAP

Though diverse, the examples of contested ills I have cited exhibit a number of core similarities. All illustrate the social and intellectual centrality of specific disease entities and the assumption that a legitimate disease is discrete and has a characteristic clinical course; perhaps equally important, behavioral and emotional symptoms are presumed to reflect an underlying mechanism. In other words, what some sociologists and social critics have for decades called "medicalization" is in practice the use of time- and place-specific vocabularies of disease entities as a tool for at once conceptualizing and managing behavior and feelings. And these disease models have ultimately to be specific and somatic if they are to find wide acceptance.

A somatic identity is perhaps most fundamental. It is no accident that today's advocates for the mentally ill state again and again that "it" is a physical ailment no different from diabetes or cancer—and no more deserving of censure or less-than-equal insurance coverage. "The brain is an organ," as a *New York Times* editorialist put it in formulaic language (June 10, 1999), "and diseases related to this organ should be treated like any other medical illness." President Clinton agreed when he described it as "morally right" for insurance companies to set the same annual and lifetime coverage limits for mental as physical ills. Like the rules of criminal responsibility, insurance coverage presents a continuing occasion for debating the nature and treatment of emotional and behavioral ills. It is not surprising that such claims inevitably generate not only novel entities but also—and equally important—somatic rationalizations for the existence of such ailments. As I prepared this chapter, for example, my morning newspaper reported that "a national study . . . reported that at some point in their lives, about 5 per cent of people have such frequent, serious blow-ups that they qualify as suffering from Intermittent Explosive Disorder, a full-fledged psychiatric diagnosis." The reporter also cited the comments of an authority on anger: "It's not simply bad behavior," the expert explained: "There's a biology and a psychology and a genetics and a neuroscience behind this, and you can come up with strategies for intervention just like for anything else, like diabetes or hypertension or depression."[10] All such examples—and one could cite hundreds of others from contemporary print and electronic sources—reflect the underlying historical reality I have already discussed: the cultural pervasiveness of somatic, mechanism-based ideas of disease specificity and the problems associated with using such concepts to manage deviance, rationalize idiosyncrasy, and explain emotional pain.

I have been interested for many years in the history of such putative disease categories,[11] and one theme that seems particularly fundamental is the idea of disease specificity itself—the notion that diseases can and should be thought of as entities existing outside their unique manifestations in particular men and women. These ideas did not become culturally pervasive until the last third of the nineteenth century, and, not so coincidentally, it was only in this period that such hypothetical disease entities began to be used widely and routinely to explain an increasing variety of socially stigmatized or self-destructive behaviors. Of course, there are earlier examples of similar phenomena. Even nonhistorians have come across such conditions as hypochondriasis, hysteria, and melancholy and of references to a variety of painful moods explained in terms of a speculative but materialist pathophysiology. Humoral explanations of temperamental peculiarity are as old as Western medicine itself, but the disease concepts they rationalized were fundamentally different from those most of us take for granted today; conditions such as melancholy or hysteria were as much flexible descriptions of individual life-course outcomes as diseases conceived of in terms of modern notions of mechanism-based specificity.

The late nineteenth century was an era of expanding clinical boundaries, a period in which hypothetical ailments such as homosexuality, kleptomania, neurasthenia, railroad spine, and anorexia were delineated—one might say put into cultural play—as disease entities. Some of these terms persist, while others have an archaic ring or altered meanings, yet all were described as novel clinical phenomena by enthusiastic late-nineteenth-century physicians. The timing is no accident. The first three quarters of the century had provided a series of intellectual building blocks, cumulatively suggesting a new emphasis on disease as discrete entity. Interest in clinical description and postmortem pathology had articulated and disseminated a lesion-based notion of disease. The late nineteenth century saw a hardening in this way of thinking, reflecting the assimilation of germ theories of infectious disease as well as a variety of findings from the laboratories of physiologists and biochemists. The gradual assimilation of the notion that specific microorganisms constituted the indispensable and determining cause of particular clinical entities seemed to endorse the specificity of infectious disease—and thus, by a kind of intellectual contagion, the notion of specific disease itself. Also supportive of such views was the growing prestige of what we have come to call the biomedical sciences—histology, biochemistry, physiology, and pharmacology. Collectively they spoke a reductionist, mechanism-oriented, and antivitalist language, providing a compelling and seemingly objective store of tools, procedures, models, and data that promised to delineate disease in newly precise, measurable—and thus portable—terms. The late-nineteenth-century vogue for heredity and evolution constituted another significant factor—linking biology and behavior, mind and body, past and present. And many of the putative behavioral ills described in the late nineteenth century were in fact seen as constitutional.[12] Alcoholism and homosexuality were prominent cases in point. Heredity seemed increasingly a determining (inexorable) force rather than one among a variety of factors interacting to determine health and disease. Like the germ theory, heredity provided many late-nineteenth-century physicians a reassuringly somatic mechanism with which to explain a variety of unsettling emotions and problematic behaviors.

EXPANDING BOUNDARIES AND THE REDUCTIONIST PROJECT, 1870–1900

Such explanations increasingly took the form of hypothetical disease entities, and neurasthenia was particularly prominent among such concepts.[13] Coined in the late 1860s by George M. Beard, a New York neurologist, the term incorporated an eclectic mixture of symptoms: depression, anxiety, compulsions and obsessions, sexual dysfunction and deviance, and fleeting aches and pains—both physical and mental. Although the concept might, in retrospect, be thought a forerunner of the twentieth-century idea of neurosis—itself a catchall description for maladaptive individual adjustments in Freudian and post-Freudian models of personality and pathogenesis—Beard rationalized his discovery in relentlessly material terms. He had no choice if it were to be taken seriously by his peers; social legitimacy presumed somatic identity. In Beard's view, the elusive and ever-shifting symptoms that characterized neurasthenia were reflections of an underlying weakness in the individual sufferer's constitutional allotment of nervous energy. "Physiology," Beard explained, "is the physics of living things; pathology is the physics of disease."[14] Although neurasthenia was characterized by feelings and behavior alone, Beard was confident that it rested on a firm, if still obscure, somatic basis. "I feel assured," he wrote in 1869, "that it [neurasthenia] will in time be substantially confirmed by microscopical and chemical examinations of those patients who died in a neurasthenic condition."[15] The postmortem pathology that had so impressively delineated—for example—the lesions of tuberculosis and Bright's disease in the first half of the nineteenth century would soon illuminate, Beard was convinced, this even more elusive but ultimately somatic condition.

This speculative somaticization of behavioral ills by physicians was, as we have seen, a medical tactic far older than Beard's Gilded Age "discovery" of neurasthenia. Hypochondria was, for example, in the words of Benjamin Rush a half century earlier, as much the result of somatic causes as any physical ailment. Rush sought to combat the judgmental—and widespread—view that such ills were "imaginary," mere self-indulgence. Hypochondria, he explained, "has unfortunately been supposed to be an imaginary disease only, and when given to the disease in question is always offensive to patients who are affected with it. It is true, it is seated in the mind; but it is as much the effect of corporeal causes as a pleurisy, or a bilious fever."[16]

Nineteenth-century physicians repetitively and formulaically referred to the brain as the organ of mind and mental illness as a product of brain disorder.[17] What might be called the assumption of an ultimately somatic pathology had never been questioned in regard to the etiology of grave and incapacitating mental ills, but it had been broadened in scope by the late nineteenth century to include a variety of putative disease pictures that Rush and his generational peers would hardly have regarded as appropriate objects for clinical attention. In Beard's era of self-referred outpatient neurology, a variety of compulsions and obsessions, emotional pain (often termed mood disorders today), and what might be called problems of identity (as in homosexuality) began to populate that novel urban space, the consulting neurologist's waiting room.

It is not surprising that would-be ailments as diverse as homosexuality, anorexia, and neurasthenia were articulated in roughly the same period, the 1860s and 1870s. All were presumed to have some somatic if not constitutional basis, and all explained behavior that seemed individually painful and dysfunctional, or socially problematic. Let me elaborate this argument with another example: that mid-nineteenth-century ailment called railroad spine, or spinal concussion, a diagnosis even less familiar today than neurasthenia. This neologism was associated with a growing mid-nineteenth-century anxiety—and lawsuits—following the era's frequent railroad accidents. It reflected as well a widespread disquietude in confronting the seemingly unnatural and feverish pace of railroad travel. The pathological concept was associated with John Erichsen, a London surgeon just as neurasthenia was associated with George Beard. Originally lectures given to students at University College Hospital in the spring of

1866, Erichsen's *On Railway and Other Injuries of the Nervous System* soon became a standard reference.[18] Like Beard's formulation, Erichsen's diagnostic neologism reflected a more widespread sense of cultural uncertainty. Even before Erichsen's work, the *Lancet* had editorialized about the neurological sequelae of railroad trauma.

> These symptoms are manifested through the nervous system chiefly, or through those physical conditions which depend upon the perfect physiological balance of the nerve-forces for their exact fulfillment. They vary . . . from simple irritability, restlessness and malaise after long journeys up to a condition of gradually supervening paralysis, which tells of the insidious disease of the brain or spinal cord, such as . . . follows on violent shocks or injuries to the nervous centres. These latter are the symptoms which frequently ensue from the vehement jolts and buffetings endured during a railroad collision.[19]

One can discern a rough filiation among neurasthenia, spinal concussion, soldier's heart, and shellshock in a clinical tradition that linked particular clusters of emotional and behavioral symptoms with a parallel dependence on a legitimating if, in retrospect, hypothetical physical mechanism. All also served in some measure as occasion and vehicle for social comment. It is no accident that in 1881 George Beard wrote a much-cited book called *American Nervousness* and later another entitled *Sexual Neurasthenia;* both addressed widespread cultural anxieties about self and society.[20]

Many of these novel late-nineteenth-century entities such as sexual inversion (homosexuality) and neurasthenia were soon widely adopted and cited but nevertheless remained controversial—to some clinicians and intellectuals real diseases, but to others mere self-indulgence or symptoms of a larger cultural decay. Diseases were deployed as rhetorical weapons in recurrent battles over cultural values and social practices. "Overstress," for example, was a condition noticed in late-nineteenth-century secondary schools and attributed to the urban middle class's relentless competitiveness, while sterility and hysteria could be seen as the inevitable cost incurred by higher education for women. An urban, technology-dependent—and thus unnatural—life could be stigmatized as psychically, as well as physically, pathogenic.

This all seems neat and tidy—with disease concepts mirroring and mediating both cultural angst and a widespread faith in the explanatory power of disease models. The dots are nicely connected, but the story on the ground is rather more complicated. These pathologizing tactics were neither universal nor consistently accepted in late-nineteenth- and early-twentieth-century America. While many were attracted by the certainties of somatic and reductionist styles of explaining sickness and health, others found this approach less than congenial. Christian Science, consistently enough, along with spiritualism, Seventh Day Adventism, and the Emmanuel Movement all developed in their several ways in tension with this polarizing development: at one pole a way of thinking about behavior reducible to somatic mechanisms (with an implied deterministic understanding of behavior), at the other a holistic, spiritual framework emphasizing the impact of faith and agency on health outcomes. One might think of them as two rather different styles of reductionism. Of course, many Americans—and not only lawyers—who shared a general faith in the progress of scientific understanding remained skeptical of the legitimacy and exculpatory implications of such would-be ills as alcoholism, kleptomania, anorexia, and nervous exhaustion; a parallel aura of ambiguity and disdain surrounded the older, but still current, diagnosis of hysteria. These behavioral and emotional ills and their presumed social causes echoed morally resonant controversies over class, appropriate gender behavior, and a variety of other issues—clashes in an endless cultural war in which we still struggle over the legitimacy of ills such as chronic fatigue syndrome, anorexia, fetal alcohol syndrome, alcoholism, and homosexuality without arriving at a stable consensus.

THE MORE THINGS CHANGE

Some aspects of medicine's social history have changed dramatically in the past century, others comparatively little. One that has changed little is the mediating role of psychiatry. Medicine in general and psychiatry in particular remain boundary managers—border police, examining and certifying transit documents in an unceasing battle over depression and anxiety, sexuality, and addiction. Psychiatry remains the peculiar legatee of such problems, an obligate participant in every generation's particular cultural negotiations—a kind of canary at the pitface of cultural strife. It is by no means the only player. Civil and criminal courts, welfare officers, media commentators, a variety of other specialists—not to mention patients and families—all play a role.

The search for somatic mechanisms with which to legitimate behavioral ills seems, in retrospect, a parallel and logically related continuity. The twentieth-century psychodynamic tradition, with its emphasis on family setting and individual psychological development and associated talk therapies, seems almost a kind of byway in relation to mainstream medicine, an oppositional—if culturally significant—counterpoint to a consistently dominant reductionism. Even at the height of its influence (from the 1940s through the 1970s) psychodynamic explanations of behavior and emotions remained in an uneasy and even marginal relationship to much of mainstream medicine (despite the widespread influence of such ideas outside the profession). That very marginality helps explain the recurrent attraction of intrusive somatic therapies in twentieth-century psychiatry.[21]

The dominance of reductionist styles has a long history in the explanation of human behavior—as we have seen in the hypothetical brain pathologies of the Pinel and Rush era two centuries ago—but it has an extraordinarily salient place today. We have never been more infatuated with visions of molecular and neurochemical—ultimately genetic—truth. "We're now at the point where we can begin articulating the physical basis of some of the mysterious brain functions that exist," widely read science writer Nicholas Wade explained, "learning, memory, and emotion. . . . We're at a point where we can move miraculously from molecule to mind."[22] In the not-too-distant past we have seen claims for the discovery of genetic bases for dyslexia, obesity, risk taking, homosexuality, even aggression. Many of us can remember the widespread discussion of chromosomal explanations for criminality. Today's fashionable evolutionary psychology adds a metahistorical style of biological reductionism to our culturally available store of mechanism-oriented and determinist explanations for behavioral and emotional pathologies (as well, of course, as the "normal").

But there remains a historical irony. We are in a moment of peculiar and revealing paradox—a complex and structured mix of reductionist hopes and widespread criticism of such sanguine assumptions. As a culture, we are relentlessly reductionist in presuming somatic (and ultimately genetic) causation for behavior, yet at the same time reflexive, critical, and relativist in our approach to existing disease classifications and therapeutic modalities. We have never been more aware of the arbitrary and constructed quality of psychiatric diagnoses, yet we have never been more dependent on them than now, in an era characterized by the increasingly bureaucratic management of health care and an increasingly pervasive reductionism in the explanation of normal, as well as pathological, behavior. I need only underline the way in which the *DSM* has evolved from its originally modest format—a little over 100 pages, spiral bound, with a soft cover in 1952 and 1968—to today's ponderous, more than 900-page octavo (with its numerous epitomes, visual aids, and commentaries), while wry commentators, lay and professional, scoff at the seemingly transparent arbitrariness of its categories.[23]

This inconsistency struck me, for example, with particular force in reading *Girl, Interrupted,* Susanna Kaysen's 1993 memoir of her late-1960s inpatient stay in McLean Hospital.[24] The book includes a revealing section in which the author is seated in her "corner Cambridge bookstore" reading *DSM III* and deconstructing the substance and language of the borderline personality disorder diagnosis that had justified her treatment almost thirty years before. She underlines the arbitrariness, the gender stereotypes, and the social control built into the seemingly objective language of clinical description. Kaysen's agnostic point of view reflects and incorporates three decades of political, epistemological, and feminist criticism of psychiatric nosology; there has never been a more skeptical and reflexive period. Explicit and fundamental criticism of psychiatric nosology has in fact been widespread for half a century; one need only cite the works of Thomas Szasz, R. D. Laing, and a variety of feminist and sociological critiques of psychiatric authority and the epistemology that justified it.

The paradoxical reality of such fundamental skepticism coexisting with a triumphalist reductionism is exemplified as well in the current debate over the use of psychoactive drugs. The widespread prescribing of such drugs implies and has in the past fifty years helped legitimate the specific-entity idea; bipolar disease is what responded to lithium; depression is legitimated ontologically by the drugs that treat it. But as current controversies over Ritalin and a variety of antidepressants and antipsychotic drugs, for example, suggest, these relationships have not resolved but have simply constituted a new site and designated players for the contestation of social values. Who would have guessed a generation ago that an American president would choose an issue such as the pediatric use of psychoactive drugs as a public issue as Clinton did during his presidency? Or that we would accept with barely a second thought such public contestation of a seemingly clinical problem?[25] Just as attention deficit hyperactivity disorder was, for example, being widely discussed and accepted, it stimulated—through a kind of cultural dialectic—a variety of forceful rejections of such categories as arbitrary social constructions. It was not just that children and increasingly adults "are too casually offered stimulants like Ritalin," a letter to the *New York Times* charged almost a decade ago, "but that biological reductionism lies behind the tendency to ignore the deeper social, psychological and cultural issues . . . in favor of assuming there is a disease located within their heads."[26] Perhaps such critical sentiments are in a minority, but they have been widely and articulately voiced in the past decade—to little effect.[27]

Such dissent is, of course, neither simply a technical (pharmacological) difference nor a problem of diagnostic precision or pharmaceutical industry marketing strategies. No quantity of well-conceived epidemiological studies will bring consensus in regard to children exhibiting a problematic restlessness; its lack is at some level a problem of human diversity, of social class, of gender, and of bureaucratic practice. Clinical epidemiological studies play a role, but as only one voice in a complicated and discordant conversation. Concepts such as hyperactivity are meaningful only in specific contexts. Even if the most extreme and intractable behaviors are ultimately products of still-undeciphered but ultimately specifiable genetic and neurochemical mechanisms, their social evaluation remains contingent and a subject of inevitable contestation. What are appropriate levels of attention? Of hyperactivity? What is normal and what is, in fact, being measured? When does therapeutics stop and enhancement begin?[28] The terms *hyperactive* or *attention deficit* are context dependent by definition—reflections of specific institutional realities and cultural needs. And one of those needs, as I have suggested, is the recourse to medical personnel, authority, and conceptual categories as at once legitimation of and a framework for the institutional management and cultural framing of awkward social realities.

Similar judgments could be made in regard to a variety of other multicausal and nonspecific ills. A phenomenon such as fetal alcohol syndrome, for example, might be thought of as a statistically configured point on a spectrum of behaviors and seemingly linked physical characteristics— perhaps reflecting an underlying biological substrate. Even if we can define and defend such a core as constituting a usefully predictive entity, that presumably gestational phenomenon—fetal alcohol syndrome—would still serve as only one element in a more complex and multidimensional social reality. A disease entity so defined would incorporate not only the ideal typical core of presumed victims of fetal alcohol syndrome but also all the effects surrounding it. Like the effects of a stone dropped into a body of water, the ripples are real indeed, ranging from labels on alcoholic beverages to individual guilt and anxiety to pleas for reduced responsibility in criminal justice contexts—or, as we have seen, expanded responsibility placed on alcohol-addicted mothers.[29]

Depression constitutes a parallel and even more pervasive phenomenon. What we call major depression may have a biochemical substrate, but the relationship between etiology and individual clinical outcome remains obscure. How do we balance the determined and the contingent, the genetically given and the situationally negotiated? And what is the relationship between such disabling ills and the spectrum of emotional states we casually term depression? What is the gradient of that slippery slope from temperament and situational reaction to something rather different and categorically pathological? Lay people today often put that functional distinction in linguistic terms when they say a person has a "clinical depression," presumably an extreme point on the necessary spectrum of human pain and varieties of mood.

Despite such indeterminacy, our repertoire of specific entities constitutes a powerful reality, providing a resource for individuals in thinking about themselves and for society in conceptualizing behaviors—as is indicated in the varied histories of such current and obsolete entities as hysteria, hypochondria, hypoglycemia, chronic fatigue syndrome, Gulf War syndrome, or gender identity disorder. There are scores of such problematic ills. The very social utility of these categories implies their contestation. There are always winners and losers in the negotiations surrounding the attribution of such diagnoses; the social legitimacy—and often social resources—associated with the sickness role constitutes a prize worth contesting. Advocacy groups and the Internet have only exacerbated such debates as some individuals claim the sick role's legitimacy offered by certain controversial disease entities—chronic fatigue syndrome, for example—while others scorn them as mere excuses for self-indulgence.

CONFLICT AND CONTINUITY

I have tried to describe a phenomenon that is always in process, always contested, and never completed. Sociologists and historians have described the linked phenomena of medicalization and bureaucracy as having mounted a powerful campaign for cultural and institutional authority over problematic behaviors and suspect emotions. And, in fact, the boundaries of presumed disease have in general expanded relentlessly in the past century and a half, but these boundaries remain contested even as they move outward. At least some medical and lay hearts and minds remain only partially converted to these new and expansive models of pathology.

This is only to have been expected. There are a number of continuities that guarantee both the continued centrality *and* contestation of behavioral and emotional ills. One such continuity turns on the paradox of using reductionist means for holistic—cultural—ends. As disease definitions have become more and more dependent on seemingly objective signs (first physical diagnosis, then laboratory findings and imaging results), ailments that cannot easily be associated with such findings are naturally segregated into a lesser status. Behavioral ills thus fall into a lowly position in a status hierarchy that is at once social, moral, medical, and epistemological. When allied with the fear, punitiveness, hostility, stigmatization, personal guilt, and pain often associated with such contested behaviors, it is hardly surprising that individuals exhibiting emotional and behavioral "symptoms" would not be consistently well served by the mechanism-oriented specific-entity style of legitimating and conceptualizing disease. And when there is no cultural consensus—as in regard to homosexuality or substance abuse—there is no basis for a nosological consensus. It is, however, equally, if ironically, inevitable that the powerful concept of disease specificity has been—and will continue to be—employed as a tool for the ideological management of problematic emotions and behaviors. It is a tool, moreover, available to laypeople as well as clinicians and administrators; there is always an eager market for disease labels, whether found on a website, a magazine, or a nosological table. Insofar as our ideal-typical conception of disease is specific and mechanism-based, this reductionist model will remain to a degree inconsistent with the cultural and bureaucratic work performed through the articulation and deployment of such disease categories.[30]

A second factor is a never-ending negotiation over agency and responsibility. Post-nineteenth-century models of disease bear with them an aura of determinism and can thus have a potentially significant role in shaping the social role allotted the sick. We want moral meaning in the narratives we impose on ourselves and others, and it is difficult to find it in the random nemesis of genetics and neurochemistry. It is not only in the courtroom but also in society more generally that we seek to preserve responsibility for individual decisions—and thus meaning in misfortune. Contemporary debates over "obesity" represent an example of such ambiguity; is it a disease or a failure of character? Does "it" represent the working out of genetic destiny, or does a predisposition toward weight gain constitute just one aspect of a complex and poorly understood biological and psychosocial identity? When does idiosyncrasy become pathology?

A third factor both nurturing the use of lexicons of diseases and—at the same time—guaranteeing conflict about their definition and legitimacy are the linkages that structure medicine into a bureaucratized and highly institutionalized society. Each diagnosis links an individual to a network of bureaucratic relationships and often specialty practice; if it can't be coded, as the saying goes, it doesn't exist. But those coding decisions are potential sites of social contestation—in which the legitimacy of individual diagnoses can become structured points of conflict and contestation. Linkage means connections, but differing institutional interests and practices breed conflict over policy, authority, and jurisdiction. I refer, for example, to debates over workers' compensation or product liability, as well as to the more obvious questions relating to disability or criminal responsibility. And, of course, individuals do not track neatly on to generalized disease categories and related practice guidelines; the potential arbitrariness of such clinical realities is often apparent to both physician and patient.

We may consign certain feelings and behaviors to the sphere of medicine, but medicine itself is not clearly bounded. Government policies on health-care reimbursement, for example, or the regulatory procedures of the Food and Drug Administration—like the often related corporate decisions of pharmaceutical companies in the private sector—have, in their various ways, shaped disease definitions, accepted therapeutics, and individual experience. Powerful stakeholders are involved in all these decisions, and all relate ultimately to the clinical practices and legitimating concepts of contemporary medicine—nowhere more markedly than in psychiatry. Consumer advertising, as well as randomized clinical trials, figure in the creation and diffusion of hypothetical disease entities, but the process is complex and elusive. Despite the expenditure of millions of advertising dollars, it is not clear, for example, that erectile dysfunction has been accepted as a legitimate, value-free, disease entity; it is still surrounded by a penumbra of stigma, whimsy, and self-conscious cultural irony.

Fourth, psychiatry and its concepts bleed constantly and unavoidably into the larger culture. This is a phenomenon by no means limited to the past century. I need only refer to the linguistic archaeology of once-technical terms adapted into everyday discourse: nostalgia, hypochondria, sanguine, hysterical, paranoid, narcissism, degenerate, nymphomaniac, psychopath, inferiority complex, obsessive compulsive. Usages have changed, but the process remains straightforward enough—ordinary men and women appropriate once-technical medical language and explanatory frameworks to think about human behavior and its social management. Behavioral and emotional ills seem more accessible than "somatic" ills to laypeople who often question such categories as depression or attention deficit but rarely interrogate—and are generally unaware of—the indeterminacy built into the diagnosis or staging of a somatic ill such as cancer.

Thirty years ago—coincidentally—I wrote an essay called "The Crisis in Psychiatric Legitimacy." In it I emphasized the difficult role played by psychiatrists and suggested that it would continue to be ambiguous no matter what technical progress might take place. "Unless all psychiatry should thaw, melt, and resolve itself into applied pharmacology there seems little possibility of these difficulties redefining themselves"[31] Perhaps psychiatry *has* in good measure resolved itself into applied pharmacology in the past three decades, but the range of human dilemmas that we ask medicine to address has if anything expanded, from depression to anxiety, from bereavement to dysfunctional marriage. As long as medicine in general and psychiatry in particular remains our designated manager of such problems, specific disease categories will always be an indispensable tool in the performance of that social role. As long as we ask medicine to help in doing the cultural work of defining the normal and providing a context and meaning for emotional pain, we will continue to fight a guerilla warfare on the permanently contested if ever-shifting boundary dividing disease and deviance, feeling and symptom, the random and the determined, the stigmatized and the deserving of sympathy.

NOTES

I would like to thank audiences at the University of Alabama–Birmingham, Ohio State, Cornell (Department of Psychiatry), Duke, Harvard, the Massachusetts Institute of Technology, the University of Pennsylvania, and Rutgers, who tolerated and provided helpful reactions to earlier versions of this chapter, and to Robert Aronowitz, Charles Bosk, Drew Faust, Gerald N. Grob, Anne Harrington, David Healy, Arthur Kleinman, David Mechanic, and Rosemary Stevens, who also read and commented on drafts.

1. Julian E. Barnes, "Insanity Defense Fails for Man Who Threw Woman onto Track," *New York Times,* March 23, 2000. Goldstein's conviction was overturned by the New York State Court of Appeals in 2005. Anemona Hartocollis, "Court Overturns Murder Conviction of Man Who Pushed Woman onto Subway Tracks," *New York Times,* December 21, 2005.

2. "Trial in Case of Drowned Children Opens," *New York Times,* February 2, 2002. Angela Yates was found not guilty by reason of insanity in a second trial. Angela K. Brown, "Yates Found Not Guilty Due to Insanity in 5 Drownings," *Boston Globe,* July 27, 2006.

3. For recent sociological overviews of the state of play, see Peter Conrad, "The Shifting Engines of Medicalization," *Journal of Health and Social Behavior* 46 (2005): 3–14; Adele E. Clarke, Jennifer Fishman, Jennifer Fosket, Laura Mamo, and Janet Shim, "Biomedicalization: Technoscientific Transformations of Health, Illness, and U.S. Biomedicine," *American Sociological Review* 68 (2003): 161–94; and Allan V. Horwitz, *Creating Mental Illness* (Chicago: University of Chicago Press, 2002). This chapter is not about the idea of medicalization and its history, but I do want to express a word of caution about the tendency to conceptualize "medicalization" as a reified, monolithic, and inexorable *thing*—a point of view that obscures the complex, multidimensional, and inconsistent nature of the way in which medical concepts and practices have laid claim to larger realms of social action and authority. Conrad, for example, refers in the article cited above to the role of pharmaceutical companies, consumers, and managed care as "engines that are driving . . . [the] medicalization train . . . into the twenty-first century" (Conrad, "Shifting Engines of Medicalization," 12). Trains are material things that move forward in one direction only—along predetermined tracks.

4. *Boston Globe,* February 20, 2001.

5. Ronald Bayer, *Homosexuality and American Psychiatry: The Politics of Diagnosis, with a New Afterword on AIDS and Homosexuality* (Princeton, N.J.: Princeton University Press, 1987).

6. David Healy has been particularly influential in his linkage of pharmaceutical company strategies with a shift in psychiatric nosology and clinical practice. See his *The Anti-Depressant Era* (Cambridge, Mass.: Harvard University Press, 1997).

7. Many of these generalizations might also be seen as applying to chronic disease, in which the stigmatization of cultural deviance is transformed into the—seemingly—neutral language of risk and in which agency and the discussion of behavior in the form of lifestyle management becomes central. See "Banishing Risk," chapter 4 in this volume.

8. Gerald N. Grob, "Psychiatry's Holy Grail: The Search for the Mechanisms of Mental Illness," *Bulletin of the History of Medicine* 72 (1998): 189–219.

9. "The work done by the alienist cannot remain long in the condition in which it is at present and still be considered worthy of respect by members of other branches of the medical profession," as one advocate for the specialty put it in 1902. Not surprisingly he called not only for a parity in the treatment of mental illness with that available for typhoid fever or pneumonia sufferers—but also for investment in high-status physical chemistry and a sharpening in the alienist's embarrassingly vague diagnostic categories. After all, psychiatry was presented with problems "involving . . . all questions for the preservation and continuance of the normal mental activities in a community." From Stewart Paton, "Recent Advances in Psychiatry and Their Relation to Internal Medicine," *American Journal of Insanity* 58 (1902): 433–42, quoted at 434 and 442.

10. Carey Goldberg, "Out of Control Anger," *Boston Globe*, August 9, 2005.

11. See Charles E. Rosenberg, *Explaining Epidemics and Other Studies in the History of* Medicine (Cambridge: Cambridge University Press, 1992) and "What Is Disease? In Memory of Owsei Temkin," *Bulletin of the History of Medicine* 77 (2003): 491–505.

12. This instance implies a complex relationship between the notion of specific disease and that of patterned deviant behavior as a determined outcome of a general constitutional makeup. Fashionable degeneration theory provided a framework .for explaining such phenomena. See Daniel Pick, *Faces of Degeneration: A European Disorder, c. 1848–c. 1918* (Cambridge: Cambridge University Press, 1989).

13. Marijke Gijswijt-Hofstra and Roy Porter, eds., *Cultures of Neurasthenia, from Beard to the First World War* (Amsterdam: Rodopi, 2001); Charles E. Rosenberg, "The Place of George M. Beard in Nineteenth-Century Psychiatry," *Bulletin of the History of Medicine* 36 (1962): 245–59; Barbara Sicherman, "The Uses of Diagnosis: Doctors, Patients, and Neurasthenia," *Journal of the History of Medicine* 32 (1977): 33–54.

14. George M. Beard, *Sexual Neurasthenia (Nervous Exhaustion): Its Hygiene, Causes, Symptoms and Treatment,* posthumous manuscript edited by A. D. Rockwell (New York: E. B. Treat, 1884), 15.

15. Beard, *Sexual Neurasthenia,* 217.

16. Benjamin Rush, *Medical Inquiries and Observations upon the Diseases of the Mind* (Philadelphia: Kimber & Richardson, 1812), 75.

17. Which is not to obscure the early-nineteenth-century physician's assumption that moral—emotional—causes could, over time, bring about somatic change. Mind and body were continuously and necessarily interactive.

18. John Eric Erichsen, *On Railway and Other Injuries of the Nervous System* (London: Walton and Mabry, 1866).

19. Ralph Harrington, "The Railway Accident: Trains, Trauma, and Technological Crises in Nineteenth-Century Britain," in *Traumatic Pasts: History, Psychiatry, and Trauma in the Modern Age, 1870–1930,* ed. Mark S. Micale and Paul Lerner (Cambridge, Cambridge University Press, 2001), 31–56, quoted at p. 42.

20. George M. Beard, *American Nervousness: Its Causes and Consequences: A Supplement to Nervous Exhaustion (Neurasthenia)* (New York: G. P. Putnam's, 1881) and *Sexual Neurasthenia.*

21. Joel Braslow, *Mental Ills and Bodily Cures: Psychiatric Treatment in the First Half of the Twentieth Century* (Berkeley: University of California Press, 1997); Jack D. Pressman, *Last Resort: Psychosurgery and the Limits of Medicine* (Cambridge: Cambridge University Press, 1998); Elliot S. Valenstein, *Great and Desperate Cures: The Rise and Decline of Psychosurgery and Other Radical Treatments for Mental Illness* (New York: Basic Books, 1986) and *Blaming the Brain: The Truth about Drugs and Mental Health* (New York: Free Press, 1998); Andrew Scull, *Madhouse: A Tragic Tale of Megalomania and Modern Medicine* (New Haven, Conn.: Yale University Press, 2005).

22. Nicholas Wade, "The Other Secrets of the Genome," *New York Times,* February 18, 2001.

23. Stuart A. Kirk and Herb Kutchins, *The Selling of DSM: The Rhetoric of Science in Psychiatry* (New York: Aldine de Gruyter, 1992); Herb Kutchins and Stuart A. Kirk, *Making Us Crazy: DSM: The Psychiatric Bible and the Creation of Mental Disorders* (New York: Free Press, 1997).

24. Susanna Kaysen, *Girl, Interrupted* (New York: Random House, 1993).

25. Of course, psychoactive drugs are not alone in attracting public debate. One thinks of the public debates over screening mammography or hormone replacement therapy—not to mention—in the even more recent past—stem-cell research, complete with newspaper editorials, op-ed battles, full-page ads, and television coverage.

26. A. Kohn and T. Armstrong, letter to the editor, *New York Times,* September 7, 1997.

27. Richard J. De Grandpre, *Ritalin Nation: Rapid-fire Culture and the Transformation of Human Consciousness* (New York: Norton paperback edition, updated, 2000; W. W. Norton, 1999).

28. Carl Elliot, *Better Than Well: American Medicine Meets the American Dream* (New York: Norton, 2003).

29. Elizabeth M. Armstrong, *Conceiving Risk, Bearing Responsibility: Fetal Alcohol Syndrome and the Diagnosis of Moral Disorder* (Baltimore: Johns Hopkins University Press, 2003); Janet Lynne Golden, *Message in a Bottle: The Making of Fetal Alcohol Syndrome* (Cambridge, Mass.: Harvard University Press, 2005).

30. It is, of course, no easy matter to fit moods and behaviors into neat, defensible, and differentiable boxes—a point underlined in many critiques of the DSM. Ironically, moreover, the power of the specific entity focuses clinical attention on any related states that might be construed as early stages of slippery slopes along the way to a full-blown disease. Anxieties and minor depressions are thus reshaped by their presumed relationship to well-marked conditions that they may signal (and possibly constitute)—like hypertension or elevated cholesterol levels in cardiovascular disease.

31. Charles E. Rosenberg, "The Crisis in Psychiatric Legitimacy: Reflections on Psychiatry, Medicine, and Public Policy," in *American Psychiatry, Past, Present, and Future: Papers Presented on the Occasion of the 200th Anniversary of the Establishment of the First State-Supported Mental Hospital in America*, ed. George Kriegman, Robert D. Gardner, and D. Wilfred Abse (Charlottesville: University Press of Virginia, 1975), 147.

-- --

CHAPTER 4

BANISHING RISK

Or, the More Things Change, the More
They Remain the Same

WE HONOR RANDOMNESS IN THE ABSTRACT, but seek to manage it in practice, to constrain misfortune in reassuring frameworks of meaning.[1] We want health to make predictive sense, to be based on coherent relationships between behavior and its consequences. Notions about the causation and nature of disease are—and have been throughout history— inextricably bound up with meaning and identity. Blame, guilt, and anxiety can be harnessed in powerful conjunction—as can the presumed interaction between body and mind. We are what we have done or neglected to do.

These ideas have been articulated countless times and in a variety of forms, from classical antiquity to the present. Throughout the history of Western medical thought, chronic sickness—and predisposition to acute and epidemic ills—was generally understood to be a cumulative product of the longtime interaction between a biologically unique individual and a particular environment. As the body moved through time, it required food and water, sleep and exercise; one was always becoming and, thus, always at risk. Every circumstance of life and each day-to-day decision were physiologically meaningful. Habits of living once established could lead cumulatively—but with ultimate inexorability—to sickness and death, just as they could, if properly regulated, maintain health well into old age.

It must be recalled that until the mid-nineteenth century the concept of specific disease entities was not understood in the modern sense; a cold could shade into tuberculosis, a bruise into cancer, disorderly eating habits into gout or diabetes. In this sense, a bad habit indulged in over time was, literally, the first stage in a disease process. Lack of disease specificity implied an elusive yet omnipresent nemesis, but one that could be understood, anticipated, and averted. Logically enough, seventeenth- and eighteenth-century guides to health and longevity emphasized the need to control all those aspects of life a prudent man or woman could control: diet, exercise, sleep, the evacuations, and emotions. In the terminology of the day, and following a tradition that could be traced back to Galen, such factors were termed the "non-naturals," as opposed to the "naturals," those innate factors which might also lead to disease. "By the term 'non-naturals' were understood all those things which are *essential to life,* but which *neither enter into the composition of the animal oeconomy, nor form part of the living body.* These comprehend *air, foods and drinks, motion and rest, sleep and wakefulness, secretions, excretions, and retentions, mental emotions, clothing, bathing, &c."*

It is obvious that such concern with daily routine provided an occasion for enforcing a society's behavioral norms; there could be no practical distinction between the realms of morality, meaning, and mechanism. The symptoms of moral sickness—sexual promiscuity, gluttony, sloth, uncontrolled emotional excess—inevitably undermined physical health.

Much of this age-old emphasis on regimen will seem enlightened and even prescient to health-conscious late-twentieth-century readers. But too much has intervened between the late eighteenth century and the present. It is no longer possible to share the assumptions of traditional medicine—even if particular elements in that configuration of ideas seem familiar. I would like to emphasize four themes fundamental to this traditional way of understanding health and disease, themes configured so as to constitute a way of thinking about the body antithetical to certain dominant trends in twentieth-century medicine. First was its aggregate, inclusive, and cumulative quality. In this sense, all categorical distinctions between body and mind, physiology and morality, were arbitrary, because each was formed and in that process transmuted by the countless ongoing interactions that constituted individuality. Second was an unquestioned emphasis on idiosyncrasy, not only the individuality of physiological response to drugs and patterning of symptoms in disease but also the uniqueness of every man or woman's innate constitution and its particular history since conception.[2] Third is the physiological centrality of unceasing interactions between body and mind and of the way in which this view of mind incorporated both the treacherous—and potentially pathogenic—passions, as well as those "higher" elements involved in decision making. Conscious decisions guided the individual in his or her passage between childhood and death; every aspect of day-to-day life implied choice, and thus volition. Following logically is the fourth element I want to underline—namely, individual responsibility for continued health and, in sickness, the management of recovery, especially in chronic illness. This cluster of ideas can be seen as maintaining a prominent role for volition and responsibility and thus remaining consistently relevant to the management of randomness.

They also constituted a doctrine best suited to the life choices of those who could exercise such choice; contemporaries were well aware of the inequalities mandated by the realities of class and work. A laborer could not easily vary his diet or those of his wife and children; he could not improve his health through regular horseback riding or sea voyages to the Adriatic or West Indies. Nevertheless, then as now, physicians differed, and intellectual positions blurred as contemporaries blamed, as well as excused, the poor for the sickness that so often followed them through life.

"THIS LONG DISEASE, MY LIFE"

Emphases on regimen and lifestyle were somewhat less evident in social constructions of acute and epidemic disease.[3] Chronic illness was a different matter. Each case constituted a unique aggregate of circumstance and responsibility, of morality and imprudence, of countless decisions (and thus behaviors) repeated over time. In chronic ills, one was not only a disease, but the disease was oneself. A bout with typhus fever was an episode, for example, not an identity. In flu, the patient suffers from but is not flu; one might be attacked by cholera, but if one survived, one was not defined by cholera in the sense that "dropsy," leprosy, diabetes, or epilepsy become aspects of subsequent identity. Acute infectious disease can also have long-term consequences, as the examples of polio or smallpox (with its capacity to disfigure and even blind) illustrate in their rather different ways.

The cure of chronic illness, like its cause, was woven gradually and inextricably into the fabric of life. Recovery or survival was forged over time; one was necessarily more responsible for managing one's own recovery in chronic than in acute ills. "People in acute diseases may sometimes be their own physicians," as the most widely read and reprinted late-eighteenth-century health manual noted, "but in the chronic, the cure must ever depend chiefly upon the patient's own endeavours."[4] Not surprisingly, medical writers of the late eighteenth and early nineteenth centuries routinely emphasized a distinction between chronic and acute illness—well aware of the very different sort of situation each created for patients, families, and practitioners.

The concept of predisposition that served to explain the selective exactions of acute disease was, as I have suggested, like chronic disease itself, a cumulative aggregate of constitution, circumstance, and regimen. In these instances, too, society articulated a rationalistically framed denial of moral randomness. The glutton, the alcoholic, the anxious and weak of spirit appeared to succumb disproportionately to yellow fever, smallpox, or cholera—as did the malnourished, the dirty, and the poor. In the case of smallpox, of course, nineteenth-century commentators could point to a culpable ignorance in the failure to employ vaccination.

Religious frameworks of meaning coexisted with, supplemented, and interpenetrated such physiological schemes, but by the end of the eighteenth century they could not stand alone. Educated lay people and physicians did tend to believe that culpable errors in behaviors—sin—brought temporal punishment, but only through mechanisms built into the human body. Excessive consumption of alcohol caused sickness, not through divine interposition, but as an unavoidable consequence of metabolizing an "unnatural" substance. What seems transparently moralistic to a late-twentieth-century sensibility seemed no more than an unmediated reflection of nature's teachings in the late eighteenth or early nineteenth centuries. The style of these speculative pathologies is, of course, more important than their specific content. They emphasized that mechanisms built into the body guaranteed that only "natural" practices and behaviors would prove consistent with health; not surprisingly, the "natural" overlapped with contemporary notions of the moral. Distilled spirits were unnatural by definition, their consumption necessarily dangerous; monogamy was natural to the species, promiscuity perilous to body and soul. Such admonitions—based upon the presumed design of living things—were everywhere, in schoolbooks and sermons as well as in medical treatises.[5]

In the past century and a half, our medical ideas have changed dramatically. We are most comfortable with ills that have a discrete, material, and well-understood basis—a basis demonstrable in the laboratory and at postmortem. Without an understanding of such a mechanism, or the presumption that it exists, we are hesitant to grant social recognition to pain and discomfort. Diagnosis can define a social role and often confers social legitimacy; without it, a sufferer may face the malingerer's demeaning and guilt-inducing status. It was inevitable that twentieth-century understandings of pathological mechanism should be made to serve didactic ends paralleling those arguments from design that loomed so prominently in the late eighteenth and early nineteenth centuries. They are simply too impressive in their explanatory power, integrated too tightly into our way of thinking about the world. They must inevitably bear the burden of meanings that transcend mechanism.

SPECIFICITY AND MECHANISM

The modern view of disease was developed in two stages. The first turned on acceptance of the idea that diseases were specific entities with characteristic clinical courses and underlying pathological bases (physiological, anatomical, or some combination of the two). The second stage in the nineteenth-century history of pathological ideas turned on a revolution in ideas about the causation of infectious disease. I refer, of course, to the germ theory. This new doctrine seemed to underline and confirm the ontological status of those entities (such as typhoid fever, for example) that had already been described by clinicians and clinically oriented pathological anatomists.

There were moral and policy implications as well. Emphasis on the existence of discrete, disease-specific, external causes implied the likelihood that disease might grow out of a chance intersection between individual and etiological agent.[6] It increasingly appeared that sickness might become severed from moral agency and thus—as some nineteenth-century commentators feared—from moral order.[7]

Individual volition and social circumstance threatened to have less and less to do with the explanation of illness. Accepting the necessity of a discrete, particulate cause emphasized the randomness of sickness as well as its mechanism; it was not *who* but *where* one was that determined vulnerability. Thus, the germ theory undercut the aggregate and cumulative aspect of traditional models of the etiology of disease (in terms, that is, of their emphasis on the prior life course of the sick person both as moral and biological individual). The dramatic discoveries of late-nineteenth-century bacteriology, therefore, gradually helped impugn the plausibility of accustomed holistic disease models and endorsed the increasing centrality of a specific, reductionist, and mechanism-oriented understanding of disease. This is a cliché of historical understanding—and like most such truisms reflects a certain measure of truth.

But not without modification. Longtime habits of mind could not be changed overnight. Environmental, constitutional, and occupational factors remained prominent in late-nineteenth-century etiological thinking, especially in explaining the incidence of chronic illness such as tuberculosis. Predisposition and susceptibility still needed to be explained, even if one conceded the necessity of microorganisms to the working out of particular disease processes. Stress, morale, and the passions continued to play a role in medical and lay thinking about health, and likelihoods of intersection with pathogenic organisms could easily be seen in social—and thus potentially policy—terms. Occupation or origin, as distinct from moral status, could site an individual in a particular social space and thus explain differential incidences of particular ills. A foundryman or metal polisher, for example, might be particularly susceptible to tuberculosis or other chest ailments, no matter what his personal habits. A poor widow might find it difficult to rent an adequately ventilated apartment, no matter how clean and regular her habits. Predisposition could be social as well as biological.

There was no easy solution to the political and moral dilemma of apportioning social and individual responsibility for the maintenance of health; late-nineteenth-century public health advocates always factored into their thinking the contributory agency of even the most economically and socially disadvantaged victims of disease. One could always find reasons to blame the impoverished and exploited, even in the course of calling for reform in their conditions of life. Workers were routinely accused of alcoholism, poor personal hygiene, and "imprudence," even when it was conceded that they lived in inhumane conditions, worked lengthy hours, and suffered periodic unemployment. The code words "intelligent" or "unintelligent" were often used to explain susceptibility—and status. The germ theory of infectious disease was to have an enormous impact, but it could not abolish such deeply felt social assumptions.

The germ theory, however, had more than intellectual consequences. It was soon built into a structure of technical qualification, laboratory findings, and professional status. This was a world of credentials and experts, increasingly (though by no means consistently) uncongenial to the holistic, individual, and moral understanding of illness that had for millennia helped men and women deal with its incursions. The germ theory was no mere academic abstraction; in the last quarter of the nineteenth and first quarter of the twentieth centuries, it would help transform every aspect of medicine.

One aspect of that transformation lay in the elevation of the profession's social status and in a recasting of the physician's social identity. We have come to expect efficacy and technical activism as appropriately characterizing the physician's ministrations and—as already emphasized—to see medical knowledge in material, mechanism-oriented, and reductionist terms. In the paradoxically linked yet contradictory elements of art and science that traditionally constituted medical acumen, the science component has become increasingly dominant. This style of framing health and disease also affects how we as patients feel about ourselves, how we construe our own sickness or continued health.

In terms of our present discussion, however, I want to underline another and perhaps less familiar aspect of twentieth-century medicine. This is the changed ecology of disease. However we weigh the relevant variables—however much credit we are willing to allot to the specific input of scientific medicine and how much to generally improved standards of living—it is undeniable that the acute infectious diseases have become far less prominent in the course of the twentieth century.[8]

As we are all aware, men and women have come to live longer in the Western world and are more likely to live with and ultimately die from chronic ills that demand social and personal adjustment and defy one-dimensional etiological explanation. Second, we have become increasingly wedded to diagnostic categories defined operationally in terms of laboratory procedures: biochemical and histological, immunological and imaging. Third, not only have we described a host of new ills, we have also turned our attention to a variety of proto- or incipient ills, the by-products of our effort to monitor and understand the normal and the pathological. I refer to such states as elevated cholesterol level or hypertension. Whatever one thinks about their ultimate prognostic or clinical significance, it is undeniable that, once delineated, such states create new emotional agendas. Individuals who harbor in their bodies the immanent menace of an elevated cholesterol or blood pressure reading face years and perhaps decades of newly meaningful decisions about every aspect of their day-to-day routines.

Twentieth-century medicine has also created new chronic disease as an unintended by-product of technical change. Therapeutic progress has turned a variety of once rapidly fatal ailments into long-term problems of living. I refer to those individuals who survive as chronic "management" problems—diabetics after the introduction of insulin or those ill with degenerative kidney disease after dialysis—and whose social and intrapsychic identities have as much as their physiological status to be managed over time. Each such case is clothed with potential moral valence. All of the gradients, choices, and consequent guilts of "lifestyle" management in those only potentially ill are exacerbated in men and women actually suffering from chronic disease. Is the patient compliant or not? How do they manage themselves? If they suffer from a chronic, yet infectious condition such as AIDS or tuberculosis, how responsible are they for seeking available treatment and in avoiding situations in which the ailment might be transmitted?

All of these changes in patterns of morbidity have helped structure a fundamental ambiguity: our general conceptions of disease have become increasingly specific while—as individuals—we have become increasingly likely to suffer from vague, multicausal and overlapping ailments. As we have shifted our burden of illness toward the chronic, so have we enlarged our store of specific clinical pictures. And the accepted repertoire of such disease categories constitutes in the social sphere a stock of archetypical narratives—trajectories against which an individual judges his or her prospects for future pain, incapacity, or death. A diagnosis may also place one's past behavior in a very different retrospective light, recasting one's sense of autobiographical self in altered terms. Every clinical entity once articulated and accepted in its several ways in the lay and medical community becomes a factor in physician-patient relations, in the patient's expectations and self-evaluation, and in intrafamily relationships. A nosology can thus be seen as constituting a culturally available menu of alternative narratives.[9] How does one arrest or modify the pathological process unfolding in one's body? How does one prevent such illness through appropriate regimen?

It is a contemporary situation subtly different from, yet paralleling, the general need for physiological prudence and foresight implied by the nonspecific pathologies of traditional medicine. Physicians have always been aware that chronic ills—no matter their original etiology—implied questions of individual adjustment and social support; time itself and increased lengths of survival imply special problems for society and for the individual patient.[10] The omnipresence of chronic disease at the end of the twentieth century (along with what I have called "protodisease states," such as elevated blood pressure or cholesterol levels) has placed increasing emotional and policy emphasis on lifestyle and individual choice— and thus responsibility. Chronic disease always implies an extended narrative, a Patient's Progress of choice and moral self-definition, just as it does a new social role. The concepts of "narrative" and "social role" express two aspects of—and ways of thinking about—the same lived reality.

It also implies a debate about the locus of responsibility. What is the role of the state in a just—and aging—society? How does one balance individual responsibility against the need for social intervention? How much does the entirely rational attempt to reduce risk through individual suasion serve to blame victims and avoid the necessity of dealing with structural inequities? This drama of contested moral choice and social commitment is one that all twentieth-century societies have acted out in their particular fashion.

BEING SICK/BECOMING SICK

Not too long ago, the U.S. Department of Health and Human Services published an elaborate position paper entitled *Healthy People 2000: National Health Promotion and Disease Prevention Objectives*, "a statement of national opportunities." Although calling for an eclectic variety of health-promoting efforts, this document nevertheless highlights the role and responsibility of individuals. As we reach the end of the century, the first page of the introduction notes, "biomedical research has available sophisticated techniques for diagnosing and intervening against disease." Medicine has taught us much about those factors predisposing to sickness and premature death "and therefore about actions that each of us can take to control our risks for disease or disability. We have learned," the document continues, "that a fuller measure of health, a better quality of life, is within our personal grasp." Smoking of cigarettes, consumption of high-fat foods, abuse of alcohol, for example, and sedentary lifestyles have all been shown to substantially increase risk of sickness and premature death. Mortality from these causes, the authors contend, "are examples of the impact of personal lifestyle choices on the health destiny of individual Americans and the future of the nation."[11] Three basic elements of a far older view of disease causation are thus linked logically, rhetorically, and emotionally: "personal," "lifestyle," and "choice." Chronic illness becomes in this moral sense an aggregate of cigarettes smoked, seatbelts unfastened, glasses filled and emptied, and cheeseburgers devoured.

As in the pre–germ theory era, habits slide gradually yet finally irreversibly into disease. Our bodies reflect cumulative behaviors and impart worth, with the specter of disease serving as sanction for accepted behavioral norms. The ideal-typical trajectories built into twentieth-century concepts of specific disease imply individual dramas of right and wrong, impulse and denial. Inevitably such physiologically informed perceptions of self fit easily and consistently into the more general cultural values of control and of the achievement of moral stature through the denial of material satisfaction. Eschewing fried foods is resonant with moral, as well as biochemical, meaning. Exercise brings a sense of worth as well as improved cardiovascular status, and monogamy imparts to believers moral stature as well as risk reduction. Cultural order and thus emotional reassurance remain implicit in a predictable relationship between willed acts and their physiological consequences—as they had in earlier centuries. Self-denial has historically been associated with spiritual stature, and this powerful association correlates nicely with much contemporary understanding of the relationships among diet, exercise, and the etiology of chronic disease.

There are certain other obvious parallels between contemporary and traditional views of chronic disease. One is the omnipresence of what I have already referred to as protodisease states, such conditions as elevated cholesterol or blood pressure levels. Both are artifacts of medicine's reductionist and laboratory-oriented style of practice, yet, ironically, each such problematic physiological status presents a potential site for moral action. Each also represents a diffusion and modification of tightly bounded conceptions of specific disease (and thus harkens back to pre-nineteenth-century conceptions of disease). In hypertension, let us say, the normal melts imperceptibly into the pathological and vice versa; boundaries are not discrete and categorical, unless we choose to create arbitrary—if bureaucratically functional—physiological or biochemical thresholds that legitimate particular diagnoses and trigger associated therapeutic procedures. With our increasing diagnostic capacities, we have provided altered narratives for millions of people who might otherwise have lived out their lives in ignorance of nemeses lurking in their bodies.

We have in the past decades also generated another, psychologically rather than physiologically or biochemically defined, type of proto-illness. I refer to such putative entities as type-A and addictive personalities. We remain—as our ancestors did—uncomfortable with monolithic etiologies that find no room for psychological and temperamental variables. And these hypothetical "types" are remarkably fluid in their construction. One might well ask whether such personality types constitute sickness, prognostic indicators, self-fulfilling moral judgment, or normal variation. Whatever their role in etiology and diagnosis, such constructs unquestionably do project cultural second thoughts: in the case of type-A personality, doubts about the unsettling kinds of behavior that so often seem to make the self-made man; in the case of the addictive personality, the fear of loss of control. They also parallel and reaffirm traditional assumptions about the necessary connection between body and mind, health and behavior—as well, of course, as the need to deal with the "passions" if health is to be maintained.

"Psychosomatic medicine" and "holistic medicine" are concepts that have become fashionable in the second half of the twentieth century. Both can be seen as reactions against the categorical claims of the mechanistic reductionist style of medical explanation that dominated the first half of the century, and both are often seen as new and advanced ways of thinking about health and disease. Both would, however, have been seen as orthodox—if not as truisms—by any eighteenth- or early-nineteenth-century physician.

In our universe of accepted disease entities, specificity implies unity of type and predictability of course. It also implies a basis in some particular pathological mechanism. As I have suggested, this cluster of assumptions constituted a fundamental shift away from pre-nineteenth-century ways of thinking about illness. A secure, mechanism-based ontological status implies social legitimacy; the diagnosis of an accepted clinical entity imparts legitimacy to behaviors—and feelings—otherwise ambiguous and can confer the benefits of sick role status in such contested instances. All ailments are not created morally equal. Despite several generations of criticisms of medicine as excessively reductionist and mechanism oriented, "functional" ills still bear a burden of moral failure, of psychic weakness or even conscious malingering. As a consequence, men and women often seek the ironic comfort of a diagnosis based on "objective"—in practice often immunological—criteria. Thus, the familiar contemporary spectacle of patients demanding a particular diagnosis, asking that physicians see in a cluster of vague, continuing, and troublesome complaints a specific and thus legitimate ailment. For example, chronic fatigue syndrome, mononucleosis, or hypoglycemia have in the recent past constituted such contested cultural goods. Lyme disease has begun to fill this social role as well, and, like many such ailments, it has mobilized lay advocacy groups dedicated to establishing the biological—and thus moral—legitimacy of their ailments. "We are working very hard to alert the public about the seriousness of the disease," the director of the New Jersey Lyme Disease Network contended in the *New York Times*. "We are also trying to convince doctors that patients are victims of Lyme disease, not Lyme hysteria."[12]

Even if such ills are often perceived in rather different terms by physicians and patients, by the educated and less educated, by women and men, they nevertheless constitute compelling ways of thinking about oneself. We must each still negotiate our own understanding of pain and incapacity, but the physician and the diagnoses he or she is empowered to bestow constitute important elements in that negotiation.

All of us accept multifactorial models of disease causation, but webs of causation (and especially seamless ones) do not always address our need for meaning—although they make intellectual and even aesthetic sense. Likelihoods and multifactorial complexes of causation do not easily fit into most people's way of thinking about themselves and family members. Nor do they fit nicely into the emotional necessities of the physician-patient relationship: men and women want to know what is to happen next in their own lives or that of family members, not in that of a statistical aggregate. It is no accident that practicing clinicians have been notoriously slow to think in terms of probability. All patients are individuals; all pain is unique. The very words "risk reduction," and their aura of impersonal distance from the coarse realities of sickness, pain, and premature death, reflect precisely a contemporary style of reassuring public discourse—while at the same time fostering guilt and the need for control in its traditional emphasis on lifestyle and regimen.

We have come, that is, full circle in this era of concern with chronic disease. A life as lived has once again become a subject of analysis and policy debate; constitution, environment, and volition are once again speculatively configured as we evaluate the etiology and prophylaxis of important ills. Contemporary debates over anorexia, AIDS, type 2 diabetes, and lung cancer, for example, all reflect in their particular ways the persistence of this explanatory style. The movement from propensity to habit to pathological mechanism has once again become a central concern of moralists as well as epidemiologists and clinicians. Body and mind, constitution and lifestyle, choice and responsibility are not easily banished from the world of pathogenesis and public health.[13] The American Institute for Preventive Medicine issued, for example, the following list of ten New Year's resolutions for 1992. The first was stress management. The others, in order, were practice safe sex; stop smoking; avoid secondhand smoke; develop a social support network; be active; control consumption of cholesterol and saturated fat; limit intake of red meat, eggs, and cheese; moderate alcohol use; and have a sense of purpose.[14]

But this is only one aspect of contemporary strategies for dealing with the prospect—and reality—of sickness, for allaying the specter of randomness. Even if we believe that a culturally homogeneous view of the body in health and sickness still existed in 1800, we can hardly entertain the notion of such a unified vision persisting into the present. Many of us do find an unambiguous, reductionist, nonresonant view of disease compelling: from this perspective, the meaning of disease reduces itself to the mechanism that underlies it, and it is a meaning necessarily discerned and defined by the physician as scientist. Emotional security thus grows out of faith in that community of physicians and scientists to whom we have willingly ceded this responsibility. Equally appealing, if to a constituency somewhat different from that content with the seemingly value-free truths of the laboratory and the computer correlation, is the logically contrary emphasis on the etiological and therapeutic power of the individual mind and emotions.

I refer to a group of self-help advocates and believers in willed healing.[15] It is comforting, for some of us at least, to believe that emotional repression can, for example, cause cancer and that mind and emotion can cure it—despite the marginal intellectual status of such beliefs in a medical world dominated by the status and reductionist views of scientific medicine. But this emphasis on the therapeutic efficacy of properly directed emotions brings with it a reciprocal social burden: the guilt of those whose disease is exacerbated by a feeling of individual failure. Critics of this view contend that it is bad enough to fall victim to cancer or other life-threatening illness without feeling that one has somehow crafted one's own susceptibility. "In a turnabout of the age-old agonized question asking why bad things happen only to good people, we are now told that—aha!—bad things happen only to good people (so repressed you know). Cancer cells are internalized anger gone on a field trip all over our bodies. Give me a break," wrote an impassioned cancer victim rejecting one such view of her ailment's etiology. "I want to face the reality of randomness in life, as well. We humans would rather accept culpability than chaos, but randomness is the law of life."[16] Despite our growing understanding of the mechanisms that constitute the biological substrate of many diseases, we can hardly lay claim to a parallel understanding of the cultural and emotional phenomena surrounding sickness and death. We will continue to impose meanings, stigmatize victims, and use the incursions of disease to sanction—and sometimes undermine—cultural norms and social policies.

Related to such negotiations is the age-old problem of apportioning responsibility between the individual and society. There is no logical need for policies aimed at the modification of individual behavior to compete for funds with efforts to deal with more general social problems. But schematic logic has little to do with it. In practice that debate continues, as we often treat individual and social policies as alternative options in a zero sum game—and as we continue our equally traditional practice of both blaming and exculpating those disproportionately at risk.

There is no simple answer. It is difficult for most of us to think about health in other than reductionist terms, yet that way of framing our prospects does not consistently satisfy our needs for meaning and certainty. Even in instances in which we can demonstrate correlations between specific risks—for example, cigarette smoking, a high-fat diet, or sexual promiscuity—objective data are inevitably clothed with subjective resonance. Simple correlations are ordinarily far from simple, as long as they concern matters of emotional import. Behavior, guilt, and responsibility are still inextricably linked. These relationships are ancient, and in some ways they have only been recast and exacerbated by the accomplishments of modern medicine. Our expectations of health and long life have increased, while the growing dominance of chronic disease makes those expectations often seem no more than ingenuous. We know too much and not enough.

NOTES

This chapter was originally prepared for the conference "Morality, Health and History" sponsored by the John D. and Catherine T. MacArthur Foundation Network on the Determinants and Consequences of Health-Promoting and Health-Damaging Behavior, Santa Fe, New Mexico, June 21–23, 1992. I am grateful for the criticism of conference participants and, in addition, wish to thank Barbara Bates, Chris Feudtner, Renee Fox, Gerald N. Grob, Steven J. Kunitz, and Irvine Loudon, who were kind enough to read and comment on it.

1. James H. Pickford, *Hygiene, or Health as Depending upon the Conditions of the Atmosphere, Foods and Drinks, Motion and Rest, Sleep and Wakefulness, Secretions, Excretions, and Retentions, Mental Emotions, Clothing, Bathing, &c.* (London: John Churchill, 1858), vii–viii.

2. It should be recalled that *constitution* meant something rather different to physicians and laymen in all centuries before the twentieth and before acceptance of a fundamental distinction between hereditary and nonhereditary attributes. In earlier eras, constitution was seen as present at conception, yet necessarily reflecting and incorporating every subsequent interaction with the environment, both in

utero and subsequently. Although disease was not regarded as absolutely transmissible from parents to child, tendencies or weaknesses could be thus inherited, and chronic diseases, such as cancer, mental illness, gout, or tuberculosis, were generally regarded as reflecting such constitutional ("diathetical") components in their etiology.

3. The density of social resonance surrounding even acute disease varies according to the ailments' character. Influenza is brief, ordinarily not life-threatening, and recovery is normally complete; not surprisingly, it is comparatively unburdened with an aura of guilt and stigma. Cholera or plague represent very different instances, although neither constituted the same sort of social phenomenon as that of a chronic illness such as epilepsy or diabetes. The phrase "this long disease, my life," is from Alexander Pope's "Epistle to Dr. Arbuthnot" and was used as an apt title in Marjorie Hope Nicolson and G. S. Rousseau, *"This Long Disease, My Life": Alexander Pope and the Sciences* (Princeton, N.J.: Princeton University Press, 1965).

4. William Buchan, *Domestic Medicine; Or, a Treatise on the Prevention and Cure of Diseases by Regimen and Simple Medicine* (Philadelphia: Thomas Dobson, 1797), xi.

5. Charles E. Rosenberg, "Catechisms of Health: The Body in the Prebellum Classroom," *Bulletin of the History of Medicine* 69 (1995): 175–97.

6. Obviously, there were certain kinds of ailment—most prominently the venereal—in which the newer knowledge of etiology seemed entirely consistent with the teachings of traditional morality. But, then again, no one had ever doubted the contagiosity of such ills, which tended—like smallpox—to be seen as unambiguously contagious by means of a specific reproducible matter.

7. Lloyd G. Stevenson, "Science Down the Drain: On the Hostility of Certain Sanitarians to Animal Experimentation, Bacteriology and Immunology," *Bulletin of the History of Medicine* 29 (1955): 1–26; Charles E. Rosenberg, "Florence Nightingale on Contagion: The Hospital as a Moral Universe," in *Explaining Epidemics and Other Studies in the History of Medicine* (Cambridge: Cambridge University Press, 1992), 90–108.

8. In referring to the weighing of variables, I refer to the so-called McKeown controversy and to a related and even more elaborate discussion in demographic and development circles of the factors relevant to the "health transition." Thomas McKeown, an English professor of social medicine, focused on the decline of tuberculosis, which was clearly marked in the second half of the nineteenth and first quarter of the twentieth centuries, well before effective chemotherapy was available. On McKeown, see for example, Thomas McKeown, *The Role of Medicine: Dream, Mirage or Nemesis?* (Princeton, N.J.: Princeton University Press, 1979); McKeown's "The Role of Medicine [A Symposium]," *Milbank Quarterly* 55 (1977): 361–428; Simon Szreter, "The Importance of Social Intervention in Britain's Mortality Decline c. 1850–1914: A Reinterpretation of the Role of Public Health," *Social History of Medicine* 1, no. 1 (1988): 1–37; and Leonard G. Wilson, "The Historical Decline of Tuberculosis in Europe and America: Its Causes and Significance," *Journal of the History of Medicine* 45 (1990): 366–396.

9. Charles E. Rosenberg, "Framing Disease: Illness, Society, and History, in *Framing Disease: Studies in Cultural History*, ed. Charles E. Rosenberg and Janet Golden (New Brunswick, N.J.: Rutgers University Press, 1992), xiii–xxvi.

10. Steven J. Peitzman, "From Dropsy to Bright's Disease to End-Stage Renal Disease," *Milbank Quarterly* 67 (1989, Suppl. 1):16–32; Barbara Bates, *Bargaining for Life: A Social History of Tuberculosis, 1876–1938* (Philadelphia: University of Pennsylvania Press, 1992).

11. The document in its entirety underlines the need to contend with social as well as individual "risk factors," but it does reflect the assumption that modifying individual behavior in an era of chronic disease (and accidents) constitutes the most direct and effective tactic for altering patterns of morbidity and mortality. *Healthy People 2000: National Health Promotion and Disease Prevention Objectives* (Washington, D.C.: U.S. Department of Health and Human Services, 1990).

12. C. Stolow, director, New Jersey Lyme Disease Network, letter to the editor, *New York Times*, June 6, 1992; Robert Aronowitz, "Lyme Disease: The Social Construction of a New Disease and Its Social Consequences," *Milbank Quarterly* 69 (1991): 79–112.

13. Charles E. Rosenberg, "Body and Mind in Nineteenth-Century Medicine: Some Clinical Origins of the Neurosis Construct," *Bulletin of the History of Medicine* 63 (1989): 185–97.

14. *Philadelphia Inquirer*, December 30, 1991, E-1.

15. Perhaps most prominent among such advocates are Norman Cousins and Bernie Siegel, the latter of whom has perpetrated the aphorism that there are no incurable diseases, only incurable patients. Norman Cousins, *Anatomy of an Illness as Perceived by the Patient: Reflections on Healing and Regeneration* (New York: Norton, 1979); Bernie S. Siegel, *Love, Medicine and Miracles: Lessons Learned about Self-Healing from a Surgeon's Experience with Exceptional Patients* (New York: Harper & Row, 1986).

16. Barbara B. Sigmund, "I Didn't Give Myself Cancer," op-ed, *New York Times*, December 30, 1989.

CHAPTER 5

PATHOLOGIES OF PROGRESS

The Idea of Civilization as Risk

IT IS USUALLY HARD TO PINPOINT THE MOMENT WHEN one decides to write
a particular piece. In this case, however, it is easy to be precise. When my
daughter was in the tenth grade, she brought home an article that had
been assigned by her history teacher; appearing originally in *Discovery* magazine and titled "The Worst Mistake in the History of the Human Race,"
it argued that the adoption of agriculture "was in many ways a catastrophe from which we have never recovered."[1] The author marshaled ar-
chaeological and paleopathological evidence to contend that the transition
to settled agricultural life brought malnutrition, crowding, and the do-
mestication of endemic and epidemic disease. The linked cultural and biological history of humankind had not, contrary to what so many of us
have confidently assumed, been a tale of linear progress toward our present enviable health status.

We have all encountered versions of this argument—and, in particular, the notion that the incidence of much late-twentieth-century chronic
disease reflects a poor fit between modern styles of life and humankind's
genetic heritage as shaped in countless centuries of hunting and gather-
ing. The notion is encapsulated in the title of another programmatic essay I have used myself in teaching: "Stone Agers in the Fast Lane." The
Late Paleolithic era, its authors argue, "may be considered the last time pe-

riod during which the collective human gene pool interacted with bio-environmental circumstances typical of those for which it had been originally selected."[2] Not surprisingly, a body evolved to fit the hunter-gatherer's life could not be expected to fare well in today's very different environment. Chronic disease, the argument follows, has developed out of this growing asymmetry—and the prevention of such ills must reflect an understanding of our organism's genetically mandated style of life.

These arguments have been widely disseminated in the 1990s. "In the past thirty years," as a *New Yorker* essay retailed a version to its readers, "the natural relationship between our bodies and our environment—a relation that was developed over thousands of years—has fallen out of balance."[3] Although they seem contemporary, such ideas immediately called to mind earlier and seemingly dissimilar research of my own—on nineteenth-century America and the perceived health dangers of urban, industrial society. My doctoral dissertation was a study of cholera in mid-century America, and, almost as soon as I began serious research, I became aware that contemporaries regarded urban life as inherently dangerous.[4] Such fears were widespread not only in this country, of course, but also in England and on the Continent: "The growth of civilization," as the London *Times* noted in 1868, "means the growth of towns and the growth of towns means, at present, a terrible sacrifice of human life. . . . The fact is that in creating towns, men create the materials for an immense hotbed of disease, and this effect can only be neutralized by extraordinary artificial precautions."[5] Endemic fevers, tuberculosis, and elevated infant mortality rates exemplified the city's everyday perils; cholera seemed no more than an acute confirmation of these grim chronic truths.

My next—immediately postdoctoral—research project on late-nineteenth-century ideas of degeneration made me aware of another widespread expression of change-oriented cultural angst—namely, a focus on neurasthenia and hysteria, both diseases putatively caused or exacerbated by the novel realities of urban industrial society. New York neurologist George M. Beard, who popularized the concept of neurasthenia as a clinical entity, in fact, saw the new disease's growing American incidence as evidence of his country's advanced cultural and technological standing.[6]

In one sense, this ironic and persistent emphasis on the role of civilization in the causation of disease is no more than a cliché, a variation of traditional primitivistic notions, endless evocations of lost worlds in which humankind had not been corrupted by wealth and artifice—all versions and reiterations of the Garden of Eden's Faustian bargain recast in epidemiological terms. Equally conventional was and is the use of disease incidence and theories of causation and pathology as vehicles for the articulation and legitimation of cultural criticism; disease has always been construed as both indicator and product of less-than-ideal social conditions.

These mid-nineteenth- and late-twentieth-century renditions of the dangers consequent upon the prideful workings of the human mind were similar in form and in their central irony, but they were also quite different, products of very different intellectual—and demographic—worlds. Both their differences and resemblances deserve analysis, and in the following pages I hope to sketch some of the nineteenth-century versions of such ventures into didactic cultural pathology, trace their conceptual descendants into the twentieth century, and then, in conclusion, specify a number of elements that have made this narrative so tenaciously plausible over time.

THE TRADITION OF RIGHT LIVING

That artificial and effete styles of life could cause disease was a sentiment far older than the nineteenth century. Traditional emphases on regimen in the preservation and restoration of health routinely incorporated a criticism of luxury and idleness. The urban merchant and his wife, the dissipated aristocrat, and the scholarly clergyman were all at risk in mind and body. Such lifestyle jeremiads were consistent with not only primitivist values and the spiritually enhancing equity of self-denial but also equally ancient—aggregate and environmentally oriented—physiological notions: that is, as the body moved through time it interacted with every aspect of its surroundings to determine health or disease. By the end of the eighteenth century, the conventions of the argument were well established. Change from savage to settled rural and then to urban life brought with it conditions increasingly inimical to the body's requirements for diet, exercise, and stable emotional surroundings. If the healthy body needed fresh air, exercise, sleep, and a moderate diet, it was also dependent on the sensations and stimuli that surrounded and impinged on it.

The body was thought of as an obligate sensation-processing mechanism, and novel or intense sensory stimuli could overload the system and cause ailments of both mind and body.

Not surprisingly, some physicians and moralists concerned with the dangers of civilization began to emphasize ailments of the nervous system. Thomas Trotter, for example, a widely read clinician and social commentator, saw English life in the first decade of the nineteenth century as causing a veritable epidemic of nervous disease; like most of his medical contemporaries, Trotter assumed that primitive peoples never suffered from such ills. "Thus health and vigour of body," he argued, "with insensibility or passive content of mind, are the inheritance of the untutored savage; and if his enjoyments are limited, his cares, his pains, and his diseases are also few." Englishmen, too, had in simpler days avoided these pathological outcomes of excessive mental work and social uncertainty. When a man turns away from his "earthen-floored straw-clad cottage, on the outskirts of the forest, he necessarily undergoes a prodigious change of circumstances. He forsakes a mode of life that had been presented to him by nature; and in adopting a new situation he becomes a creature of art."[7] Much of this change had taken place within the relatively recent past, Trotter believed; he compared Thomas Sydenham's seventeenth-century contention that fevers constituted two-thirds of mankind's ills with his own clinical experience, which demonstrated that nervous ailments had assumed the dominant place previously occupied by fevers.[8] England had become more prosperous, more urban, and thus more vulnerable to the emotional volatility of civilized life.

Concern for the psychic dangers of an artificial and emotionally fevered life had become conventional by midcentury.[9] "The mental faculties are the thinking man's tools, constantly in use," as prominent psychiatrist D. Hack Tuke explained in 1857, "and often subject to very rough usage, but still oftener to unnecessary wear and tear—their employer not unfrequently totally unaware that in producing certain results he is using any tools at all."[10] Could one doubt that civilized human beings were far more liable to mental illness than nomadic or rural peoples?

Knowledge brings with it its miseries as well as its blessings. The tree in the Garden of Eden, which was "a tree to be desired to make one wise," was nevertheless the tree of the knowledge of evil, as well as of good. . . . . Civilization, with its attendant knowledge and education, creates social conditions, and offers prizes dependent solely upon intense intellectual competition, unparalleled in any former age, and of course unknown among barbarous nations, which of necessity involves *risks* (to employ no stronger term), which otherwise would not have existed.[11]

Refined sentiments as well as the intense use of their intellect distinguished civilized from primitive peoples. "What, indeed, can be a greater contrast than that which is presented by the untutored savage, on the one hand, and the member of a civilized community, on the other?"[12]

It was equally clear to Tuke and his contemporaries that the educated and prosperous were more vulnerable to psychic stress and consequent disease than was the urban working man; the "better sort" were more exposed to the sensory input of novel ideas, economic volatility, and the anxiety-producing stimulation of religious and political choice. "The rich," as Benjamin Rush had explained it earlier in the century, "are more predisposed to madness than the poor, from their exposing a larger surface of sensibility to all its remote and exciting causes."[13] The centrality of regimen and lifestyle in traditional explanations of health and disease meant that such hygienic speculations were always vehicles for reflections on class.

These views were no more than conventional, at once occasions and rationales for culturally ubiquitous calls to reconsider and reorder modern ways of life—what I have already referred to as a lifestyle jeremiad. The argument turned on assumptions about process, on the ongoing relationship between the body and its surroundings—and it was assumed that an environment untouched by humankind's own intellectual efforts made for the most natural and, thus, the least physiologically stressful environment. Most bodily ills developed out of humankind's cognitive abilities allied with free will; they could not grow out of a natural, health-defining environment, one undisturbed by human acts: "every Thing is best as it has been," in the words of eighteenth-century moral hygienist George Cheyne, "except the Errors and Failings of our free Wills."[14] In similar terms, public health advocate Benjamin Richardson explained in his *Diseases of Modern Life*, a century and a half later, that the "perfect law" of health could be fulfilled only when "the carrying of it out is retained by Nature herself:

[when] human free-will and caprice that springs from it have no influence."[15] Artifice implied deviation from design, and thus risk—and the measure of the unnatural was the measure of that risk. Thus, for example, it was easy enough for some nineteenth-century temperance advocates to draw a distinction between distilled—and hence unnatural—alcohol and a more innocent alcohol developed in the "natural" process of fermentation.

It is hardly surprising that the connection between social change and mental illness was generally unquestioned throughout the West in the nineteenth century. It was an assumption so widely accepted that as late as 1953 two students of the epidemiology of mental illness conducted a quantitative historical study of mental hospital commitment rates, to throw light on what they described as widely accepted "contentions concerning the psychologically pathic effects of contemporary social existence . . . the conviction that 'civilization,' with its high degree of individuation, personal insecurity and competitiveness, and its killing 'pace,' is responsible for a large measure of our psychotic population."[16] They confessed that they were in fact shocked ("intimidated" is the word they used) by their conclusions, which showed no increase in the incidence of mental illness during the previous century, or in urban as opposed to rural areas.[17]

Similarly, few commentators doubted that urban population density, poor ventilation, an inadequate and adulterated diet, and the unhygienic conditions of most factories and warehouses brought fevers, tuberculosis, and summer diarrhea. That men and women lived longer and healthier lives on farms and in rural villages seemed no more than an intuitive truth—a truth confirmed by every mid-nineteenth-century student of mortality statistics. The contrast was both dramatic and instructive, and constituted a de facto blueprint and motivation for environmental reform.

SOCIAL CHANGE AND SENSORY OVERLOAD

In the latter part of the nineteenth century, however, these traditional notions presented themselves with a new urgency—and seeming novelty. They were reconfigured so as to emphasize the peculiar psychic dangers posed by an ever-more-urban, technology-based, communication-oriented, and unsettlingly mobile society. Such perceptions were exemplified by

George Beard's putative discovery of neurasthenia, a protean ailment characterized by fears and anxieties, by sexual dysfunction, vague pains, and headaches—by what Beard described quite literally as "nervous weakness." Beard claimed that he was not simply replicating traditional warnings against luxury and a sedentary life, but had identified a genuinely novel clinical picture, one produced by and indicative of a new kind of technologically based and rapidly changing—modern—society. Five characteristics of that society made massive demands, he argued, on the body's limited quantity of nervous energy: steam power, the periodical press, the telegraph, women's education, and science.[18] A variety of commentators and clinicians produced versions of similarly anxious diagnoses—and thus judgments on the society they saw around them. The human body could not, it seemed, process all the intense stimuli produced by this new society.

"The present is an age of great mental activity," prominent American alienist Isaac Ray explained in 1863. "The amount of it now required for maintaining the ordinary routine of the world would have passed all conception a century ago." Industry concentrated, diffused, and intensified such mental demands. "When we consider the amount of thought that has been concerned in bringing the manufacture of a pin or a screw to its present state of perfection, we may have a remote conception of the amount of that kind of mental exercise which is required in creating and conducting the countless processes of human industry."[19] It was no surprise, then, that mental illness was expanding relentlessly throughout the Western world.

Rapid progress in communications only added to the constant and ever-changing inputs that needed to be processed. "The humblest village inhabitant," Max Nordau argued at the end of the century, "has to-day a wider geographical horizon, more numerous and complex interests, than the prime minister of a petty, or even a second-rate state a century ago. . . . A cook receives and sends more letters than a university professor did formerly, and a petty tradesman travels more and sees more countries and people than did the reigning prince of other times."[20] The Reformation and the discovery of the Americas had changed ideas, but not material life for most Europeans. "In our times," however, Nordau argued, "on the contrary steam and electricity have turned the customs of life of every member of the civilized nations upside down. . . . Every line we read or write, every human face we see, every conversation we carry on, every scene we perceive through the window of the flying express, sets in activity our sensory nerves and our brain centres."[21] It was hardly surprising, he contended, that such novel stimuli increased suicide and mental illness. If the average man and woman found difficulty in adjusting to such intense demands, those already weakened by heredity would certainly not adjust to the demands created by this new world. Nordau is best known to cultural and medical historians for his overwrought attack on the hereditary degeneracy that he claimed suffused fin de siècle culture and Europe's urban elite. His emphasis on the place of constitutional nervous weakness not only served as a way of attacking those styles of life and artistic work that he deplored but also explained the differential susceptibility of individuals to the city's emotionally draining realities.[22]

Although civilization did create novel ailments, it must be emphasized that none of these Victorian explorers of this supposedly uncharted pathological territory advocated turning back the social clock. Beard and Nordau, for example, both felt that civilization would in its inevitable progress provide the answers to these situational dilemmas; Nordau's urban degenerates with their weakened nervous systems would probably not survive and reproduce in this new era, for example, while Beard simply assumed that technology would improve upon itself and create ever-more-viable social realities.[23] Tuke, too, emphasized on balance the positive aspects of European progress, despite the acknowledged emotional costs. He ended the article quoted above with these often-cited lines from Tennyson's *Locksley Hall:*

Not in vain the distance beacons. Forward, forward let us range,
Let the great world spin for ever down the ringing grooves of change.
Thro' the shadow of the globe we sweep into the younger day:
Better fifty years of Europe than a cycle of Cathay.[24]

And no one, of course, had ever doubted that progress in civilization would exact physical and emotional costs. The notion of an existential cost-benefit calculation suffused traditional notions of the mutually constitutive relationships among body, mind, and community; human passions had necessarily to be controlled or they would bring sickness to the individual and moral chaos to society. Civilization was, in fact, predicated upon the management of such emotions: the fear, anger, envy, and lust built into the body had always coexisted uneasily with the inhibitions demanded by an ordered, civilized society. Freud's fatalistic vision of the necessary conflict between such innate drives and the demands of civilized behavior is well known, but in some ways his views only restated and incorporated much older assumptions of an unavoidable and often pathogenic conflict between the collective demands of society and needs hardwired into the individual body.

STONE AGE MAN IN THE FAST LANE

By the end of the twentieth century, many of the same themes and anxieties had recast themselves in rather different form. Our most visible anxieties surfaced in regard to chronic disease, not neurosis or hysteria, and they have been explained increasingly and pervasively in terms not of the city as a pathogenic environment but of evolutionary and global ecological realities. If the body in 1850 was a processing mechanism always threatened by disequilibrium, the late-twentieth-century body was very much a product of evolution, an organism shaped irrevocably in deep time—and thus limited in its capacity to adapt to sharply altered environmental realities.

We have since the beginning of the twentieth century been fleshing out this image of the evolutionary past in the present. Brain cells, in the melodramatic words of the Progressive Era surgeon and speculative physiologist George Crile, have existed for millions of years, "with perhaps less alteration than the crust of the earth"; they are "bound to the entire past," he explained in 1915.[25] Crile likened the human body to a musical instrument—an organ—"the keyboard of which is composed of the various receptors, upon which the environment plays the many tunes of life; and written within ourselves in symbolic language is the history of our evolution."[26] And the nineteenth century, his synthetic argument followed, had seen a rapid acceleration in the biologic divergence that has increasingly differentiated humans from all other mammals:

As compared with the entire duration of organic evolution, man came down from his arboreal abode and assumed his new role of increased domination over the physical world but a moment ago. And now, though sitting at his desk in command of the complicated machinery of civilization, when he fears a business catastrophe his fear is manifested in the terms of his ancestral physical battle in the struggle for existence. He cannot fear intellectually, he cannot fear dispassionately, he fears with all his organs, and the same organs are stimulated and inhibited as if, instead of it being a battle of credit, of position, or of honor, it were a physical battle with teeth and claws.[27]

This style of understanding the human body in terms of mechanisms created in response to past environmental circumstances provided a whole new repertoire of speculative frames in which to understand modern civilization as risk factor. Hans Selye's widely popularized mid-twentieth-century concept of the general adaptation syndrome, for example, built on a similarly problematic mix of innate physiological mechanisms and unavoidable environmental constraints. "Life," as he put it, "is largely a process of adaptation to the circumstances in which we exist. . . . The secret of health and happiness lies in successful adjustment to the ever-changing conditions on this globe; the penalties for failure in this great process of adaptation are disease and unhappiness." Much of chronic disease, he emphasized to a general audience in 1953, had to be thought of as the consequence of failures in adaptation.[28]

Late-twentieth-century critics often pointed out the structured asymmetry between a body evolved in Paleolithic conditions and the late-twentieth-century environment in which that body maintained itself. We had, as the by-now-familiar argument explains, simply exchanged infectious for chronic disease, the triumphalist claims of medicine notwithstanding. The English social medicine advocate Thomas McKeown, for example, deployed such arguments to help underline the limits of a reductionist, acute-care-oriented medicine. Although he is best known for his emphasis on environmental factors—and especially diet—his argument proceeds logically from his foregrounding of the restraints placed by our genetic endowment on our choices in health policy. In his study of the *Origins of Human Disease,* for example, McKeown elaborated his fundamental and often-repeated distinction between what he termed diseases of affluence and diseases of poverty—basing his rationale on "the genetic stability of the human population."[29] Like many other critics of an acute-care, procedure-oriented medicine, McKeown was something of a historian as well as an epidemiologist and policy-maker. The determinants of health had changed very little before the nineteenth century, he contended,

> vast changes from the conditions under which man evolved were associated with the transitions from nomadic to agricultural to industrial ways of life. . . . Industrialization, with the advances in agriculture and increased knowledge of disease origins and mechanisms, brought relief from many of the problems associated with poverty; but by providing an excess of resources over needs, it opened the way to the ill-effects of affluence. . . . The solution of common disease problems will not, therefore occur naturally as it did during man's evolution, by elimination of deleterious genes. It is all the more important that we should understand, and so far as possible control, environmental determinants of disease.[30]

It is hardly surprising that social medicine advocates have, since the mid-nineteenth century, found such arguments congenial: the risks to health associated with material progress have served again and again as arguments for environmental reform.

Another difference between mid- and late-nineteenth-century notions of the risks associated with civilization and our own is the relatively marginal role of medicine as a substantive change agent in the earlier era, as contrasted with its importance today. Our seeming victory over infectious disease has made medicine in the popular mind both creator and inheritor of a chronic disease–centered clinical environment.[31] The laboratory has also provided a variety of therapeutic modalities—such as insulin or dialysis—that have helped reconfigure the incidence and distribution of chronic illness. In this sense, late-twentieth-century diabetes or chronic kidney disease after dialysis can be seen as pathologies of clinical progress. It is medicine, in addition, that developed and applied, and then over-prescribed, antibiotics—leading to selection for drug-resistant strains of pathogens and thus helping underline the need to integrate micro- and macrolevels of evolution in understanding the relationships among culture, environment, and disease incidence.

Popular anxieties about the pathogenic consequences of progress have shifted, as I have suggested, from a nineteenth-century focus on the city to a more expansive global perspective in which inclusive and ecological styles of analysis have become increasingly pervasive. I have, by way of example, already referred to the selection of drug-resistant bacteria in human beings, and I should add the indiscriminate use of antibiotics in America's highly industrialized agriculture to illustrate such linkages. Anxiety about such practices is by no means limited to natural food zealots or academics. Thus, for example, more than a few Americans have shown themselves willing to pay a premium for free-range, drug-free chickens as opposed to their antibiotic-doused and battery-constrained peers. It is equally natural for some of us to see the emergence of new viral diseases— as well as a variety of social pathologies—in the destruction of rain forests and the recasting of traditional economies and ways of life. These are hardly new ideas, even if AIDS and a variety of novel ills have forcefully restated them. "Man," as Macfarlane Burnet argued almost a half century ago in his much-cited *Natural History of Infectious Disease,* "lives in an environment constantly being changed by his own activities, and few of his diseases have attained . . . an equilibrium."[32] An unceasing manipulator of the environment, humankind has never been able to attain a stable ecological relationship with the universe of potential pathogens. These explanatory models, like their nineteenth-century predecessors, have, moreover, underlined the links between cultural and biological change— making the epidemiologist an obligate social historian and potential moralist as well as a student of mortality and morbidity.

And disease incidence does, in fact, seem to be associated with environmental and lifestyle factors: a variety of widely cited studies have shown dramatic change in site-specific cancer rates as men and women from less-developed economies move to industrialized societies; parallel studies have highlighted similar patterns in the incidence of hypertension.[33] The point seems obvious enough and underlines the didactic and expository usefulness of the "Stone Agers in the fast lane" metaphor. It is a powerful and politically resonant trope. Civilization, in the form of global capitalism, has become an all-pervasive and seductive force—seeming to produce AIDS along with sneakers, blue jeans, and inexpensive consumer electronics. Regardless of whether one accepts such explanatory hypotheses, all of us must reckon with this widely disseminated, multidimensional model and—

despite its epidemiological form—with its latent moral and political content. "We already know that some of the better aspects of modern life are carcinogenic," a gerontologist wrote to the *New York Times* a decade ago: "Instead of reflexively blaming pollutants, we may have to live with the fact that there is a downside to the pleasures and indulgences of existence at the end of the 20th century."[34] Similarly, a recent popular history of disease explained that its theme was that "human progress breeds disease, and has always done so," and that the book would "try to make it clear that over the course of their short stay on earth humans have helped to create their plagues, poxes and pestilence by unwittingly fashioning the kinds of circumstances that brought them forth and then, at times, almost compulsively improving those circumstances so that the diseases flourished."[35]

As the morally resonant language of these quotations demonstrates, speculations about the relationship between material change and disease incidence serve to represent and legitimate dismay about key aspects of our society—cultural anxieties that present themselves in the formally value-free terms of meta-epidemiology. These authors are clearly expressing a critical agenda; expressions such as "downside to the pleasures and indulgences" or "almost compulsively" make clear the emotional subtext informing their authors' prose. It is hardly surprising that invocations of the pathologies consequent upon progress have so often been oppositional in the late twentieth century—jeremiads that criticize dominant tendencies in medicine and in society generally.

CONCLUSION: PERSISTENT MEANINGS

Why are these rhetorical strategies so powerful? Why is this story of the ultimate ambiguity of human striving so persistent? One might argue that it does no more than reflect reality. And, in some instances, disease *does* seem a necessary consequence of social and economic growth. Death rates in mid-nineteenth-century London or Birmingham were doubtless higher than those in rural areas, and site-specific cancer rates do, in fact, change with population movements as measured by late-twentieth-century data. On the other hand, we are now less certain that mental illness and neurosis grow inevitably out of urban stress. Perhaps more important, however, the penumbra of social meanings surrounding these discussions—and that I have tried to illustrate in the authors' own words—indicates a powerful relevance to wider cultural needs and perceptions.

I would like, by way of conclusion, to suggest a number of persistent characteristics of this narrative and its subject matter that help account for its cultural tenacity. First is the very narrative character of this tale of human temerity, of progress and punishment. Like any didactic parable, it follows a trajectory that configures agency and its consequences; incident, meaning, and explanation are linked as versions of this story unfold. It is a particular kind of narrative, moreover, a moral parable underlining the ambiguous nature of human progress and of our ultimate lordship over the material domain we have presumed to rule.

Second, and relatedly, the progress-and-pathology narrative is holistic and inclusive. As I have implied, it assumes that the social system is inevitably an element in a biological system—and vice versa—and it underlines the multiplicity of interactions between mind and body, biological and moral consequence, for it is the mind's imagination, its will to alter nature, that creates cultural products inimical to the very body that houses and nourishes that inquiring mind. And this is a story, moreover, that links history with the body and its environment.

Third, the narrative is, in form, an argument from design, focusing on the structured opposition between a rightness inherent in the biologically given and an arbitrariness unavoidable in the culturally constructed. It is this very arbitrariness that creates the space in which disease is engendered. In earlier centuries, scholars and theologians had become accustomed to demonstrating evidence of a Divine Maker in the structure and function of living things. In the past century, as we are well aware, evolution rather than God has come increasingly to legitimate behavioral norms as grounded in the body's structure and functions. And, not unrelatedly, medical language and medical status have come—as I hope I have suggested in this discussion—to play a central role in this articulation, negotiation, and sanctioning of cultural norms.

Fourth, the progress-and-pathology narrative incorporates the rhetorical power of the contrast between the *is* and the *ought*. Rhetoricians of pathology have referred again and again to instructive contrasts between rates of death and disease: in the mid-nineteenth century, for example, that between urban and rural mortality rates; in the late twentieth century, contrasts in site-specific cancer rates. The lower rate is presumably a closer approximation of some natural rate built into the body—Yemenite women before their breast cancer rate soared in Israel, for example—thus underlining the role of human actions in creating a culpable gap between the natural and the artificial, the ought and the is. Such contrasts create motivation and imply policy. That which human agency has created, that same agency can recast. Thus, disease incidence becomes an argument for social reform, an indictment of a pathogenic society. It is no surprise that such arguments have proven congenial to generations of public-health activists.

Equally persuasive and persistent is the emphasis in this recurring narrative on "fit"—on the point of interaction between the body and its environment. "Adaptation" is a term we have come to use in describing both this interaction and its *successful* outcome; "stress" is the term we use in describing poorness of fit. The twentieth-century currency of terms such as "homeostasis" and "stress" mirrors in part the cultural vitality of these traditional assumptions—of the way, that is, in which a health-defining internal balance requires a stable, ongoing interaction with an appropriate external environment. It is a way of thinking about disease and health that hinges on humankind's collective ability to provide that appropriate environment—to ensure the dynamic balance that constitutes health.

It is, finally, a narrative that can be used in a variety of contexts and with a variety of social motives. Its very fluidity makes it useful to analysts and advocates of every persuasion; one might think of it as being characterized, in fact, by a kind of creative formlessness. As we have seen, reflections on the relationship between civilization and disease can be used in an oppositional way by critics of mainstream reductionist medicine— and of market relations—yet the narrative uses the tools and concepts of reductionist medicine—like disease-specificity itself—to make a holistic and antireductionist argument. It can focus on the inequalities of class or help rationalize such inequities as the outcome of differing biological endowments. Or it can be used to ignore questions of class difference by placing all men and women in a larger framework of meaning and action. It can focus on the potential role of will and agency in avoiding the dangers of a particular environment—or it can be used deterministically to highlight the individual's powerlessness in the context of structural changes; it can, that is, be used to either blame or excuse victims. It can be infused with the tones of urgent, reform-legitimating jeremiad, or—as in the case of Freud—with tragic irony. It is a parable that has been continually re-formulated and restated but always with the profound capacity to express questions about the limitations of ambitions no more than human and of progress no more than material.

NOTES

1. Jared Diamond, "The Worst Mistake in the History of the Human Race," *Discovery* 8 (May 1987): 64–66, quotation on 64. Diamond's *Guns, Germs, and Steel: The Fates of Human Societies* (New York: Norton, 1997), a more elaborately developed synthesis of his ecologically determinist views, was awarded the Pulitzer Prize for nonfiction in 1997.

2. S. Boyd Eaton, Melvin Conner, and Marjorie Shostak, "Stone Agers in the Fast Lane: Chronic Degenerative Diseases in Evolutionary Perspective," *American Journal of Medicine* 84 (1988): 739–49, quotation on 740. See S. Boyd Eaton, Marjorie Shostak, and Melvin Conner, *The Paleolithic Prescription: A Program of Diet and Exercise and a Design for Living* (New York: Harper & Row, 1988), for a lay-oriented synthesis of such views. For a parallel example of the use of ethnographic and paleopathologic data in a policy context, see also Mark Nathan Cohen, *Health and the Rise of Civilization* (New Haven, Conn.: Yale University Press, 1989).

3. Malcolm Gladwell, "Annals of Medicine: The Pima Paradox," *New Yorker*, February 2, 1998, 56.

4. The dissertation was published in book form: Charles E. Rosenberg, *The Cholera Years: The United States in 1832, 1849, and 1866* (Chicago: University of Chicago Press, 1962).

5. *Times* (London), October 8, 1868; cited in John Woodward, "Medicine and the City," in *Urban Disease and Mortality in Nineteenth-Century England,* ed. Robert Woods and John Woodward (New York: St Martin's Press, 1984), 65.

6. Charles E. Rosenberg, "The Place of George M. Beard in Nineteenth-Century Psychiatry," *Bulletin of the History of Medicine* 36 (1962): 245–59; see esp. 254, 257.

7. Thomas Trotter, *A View of the Nervous Temperament, Being a Practical Inquiry into the Increasing Prevalence, Prevention, and Treatment of Those Diseases Commonly Called Nervous, Bilious, Stomach & Liver Complaints; Indigestion; Low Spirits, Gout &c.* (Troy, N.Y.: Wright, Goodenow, & Stockwell, 1808), 23.

8. Trotter, *Nervous System Temperament,* xii–xiii. Trotter, revealingly, referred to his description of a large city's inhabitants as "a kind of medical analysis of society" (viii).

9. George Rosen, "Social Stress and Mental Disease from the Eighteenth Century to the Present: Some Origins of Social Psychiatry," *Milbank Memorial Fund Quarterly* 37 (1959): 5–32; Mark D. Altschule, "The Concept of Civilization as Social Evil in the Writings of Mid-Nineteenth-Century Psychiatrists," chap. 7 in *Roots of Mod-*

ern Psychiatry: Essays in the History of Psychiatry (New York: Grune & Stratton, 1957), 119–39.

10. Daniel H. Tuke, "Does Civilization Favour the Generation of Mental Disease?" *Journal of Mental Science* 4 (1858): 94–110, quotation on 94.

11. Tuke, "Does Civilization Favour the Generation of Mental Disease," 95. Tuke's use of the term "risk" is somewhat atypical for his generation, but its juxtaposition with his invocation of the Garden of Eden story does indicate an awareness of its moral as well as actuarial resonance.

12. Tuke, "Does Civilization Favour the Generation of Mental Disease," 97.

13. Benjamin Rush, *Medical Inquiries and Observations upon the Diseases of the Mind* (Philadelphia: Kimber & Richardson, 1812), 62. When mental illness occurred in "poor people," Rush believed, it was generally the effect of "corporeal," not moral or emotional, causes.

14. George Cheyne, *An Essay of Health and Long Life* (London: Strahan & Leake, 1724), xv. Man's moral and intellectual attributes, in the words of another moral philosopher, "while they open sources of enjoyment immeasurably exceeding any possessed by the inferior animals, beget a train of moral, and their consequent physical ills, too often filling life with sorrow, and leading almost to a doubt whether it be a gift of mercy, or an imposition of wrath." Quotation from William Sweetser, *Mental Hygiene, or an Examination of the Intellect and Passions, Designed to Illustrate Their Influence on Health and the Duration of Life* (New York: J. & H. G. Langley, 1843), 21.

15. Benjamin Ward Richardson, *Diseases of Modern Life* (New York: Appleton, 1876), 3. Death was natural—designed into the body—in Richardson's scheme, but disease was unnatural, a consequence of "the social opposition" to Nature. Richardson argued that, "if the free-will with which she has armed us were brought into accord with her designs, she would give us the riches, the beauties, the wonders of the Universe for our portion so long as we could receive and enjoy them" (5; cf. 43).

16. Herbert Goldhamer and Andrew Marshall, *Psychosis and Civilization: Two Studies in the Frequency of Mental Disease* (Glencoe, Ill.: Free Press, 1953), 21.

17. Goldhamer and Marshall, *Psychosis and Civilization*, 73. There is a rich literature debating and seeking to document such connections between social dislocation and the production of mental illness; Emile Durkheim's *Suicide: A Study in Sociology*, trans. J. A. Spaulding and G. Simpson (1897; London: Routledge & Kegan Paul, 1952), is perhaps the best known, but it was in some ways no more than representative. For some mid-twentieth-century examples, see Robert E. L. Faris and H. Warren Dunham, *Mental Disorders in Urban Areas: An Ecological Study of Schizophrenia and Other Psychoses* (New York: Hafner, 1960; orig. pub. University of Chicago Press, 1939); *Interrelations between the Social Environment and Psychiatric Disorders: Papers Presented at the 1952 Annual Conference of the Milbank Memorial Fund* (New York: Milbank Memorial Fund, 1953).

18. George M. Beard, *American Nervousness: Its Causes and Consequences* (New York: Putnam, 1881), 96. Beard distinguished between modern civilization (as marked by these characteristics) and the more generic use of the term "civilization"—as, for example, in reference to classical Greece or Rome.

19. Isaac Ray, *Mental Hygiene* (Boston: Ticknor and Fields, 1863), 224–25. "In those primitive times when a successful employment only required a certain acuteness of the senses and faculties common to man and the brutes, disease was not often induced in the brain by an undue exercise of its powers."

20. Max Nordau, *Degeneration*, trans. from the 2d German ed. (New York: Appleton, 1895), 37.

21. Nordau, *Degeneration*, 39.

22. Nordau is also known for his role in the Zionist movement. For an exposition of his ideas and influence, see Hans-Peter Soder, "Disease and Health as Contexts of Fin-de-Siècle Modernity: Max Nordau's Theory of Degeneration" (Ph.D. diss., Cornell University, 1991); and for the European contexts of use of degeneration ideas, see Daniel Pick, *Faces of Degeneration: A European Disorder, c. 1848–c. 1918* (Cambridge: Cambridge University Press, 1989). See also J. Edward Chamberlin and Sander L. Gilman, eds., *Degeneration: The Dark Side of Progress* (New York: Columbia University Press, 1985), esp. chap. 3: Robert A. Nye, "Sociology and Degeneration: The Irony of Progress," 49–71.

23. Rosenberg, "Beard," 257, n. 6. The degenerate would succumb, Nordau argued, while the great majority of the healthy "will rapidly and easily adapt themselves to the conditions which new inventions have created in humanity. . . . The end of the twentieth century, therefore, will probably see a generation to whom it will not be injurious to read a dozen square yards of newspapers daily, to be constantly called to the telephone, to be thinking simultaneously of the five continents of the world, to live half their time in a railway carriage or in a flying machine" (Nordau, *Degeneration*, 541, n. 22).

24. Tuke, "Civilization," 110, n. 11.

25. George W. Crile, *The Origin and Nature of the Emotions: Miscellaneous Papers*, ed. Amy F. Rowland (Philadelphia: Saunders, 1915), 54. "By both the positive and negative evidence we are forced to believe that the emotions are primitive instinctive reactions which represent ancestral acts; and that they therefore utilize the complicated motor mechanism which has been developed by the forces of evolution as that best adapted to fit the individual for his struggle with his environment or for procreation" (75).

26. Crile, *Origin and Nature of Emotions*, 34.

27. Crile, *Origin and Nature of Emotions*, 61–62. "Whether the cause of acute fear be moral, financial, social, or stage fright," Crile continued, "the heart beats wildly, the respirations are accelerated, perspiration is increased, there are pallor, trembling, indigestion, dry mouth, etc. The phenomena are those which accompany physical exertion in self-defense or escape" (62). Crile's argument parallels the more widely

known exposition of Walter B. Cannon: *Bodily Changes in Pain, Hunger, Fear and Rage: An Account into the Function of Emotional Excitement* (New York: Appleton, 1915).

28. Hans Selye, *The Stress of Life* (New York: McGraw-Hill, 1956), vii–viii. "For instance, we are just beginning to see that many common diseases are largely due to errors in our adaptive response to stress, rather than to direct damage by germs, poisons, or other external agents. In this sense, many nervous and emotional disturbances, high blood pressure, gastric and duodenal ulcers, certain types of rheumatic, allergic, cardiovascular, and renal diseases appear to be essentially diseases of adaptation" (viii). See also 204–5 and 276–77.

29. Thomas McKeown, *The Origins of Human Disease* (Oxford: Basil Blackwell, 1988), 141.

30. Thomas McKeown, *The Role of Medicine: Dream, Mirage, or Nemesis?* (Princeton, N.J.: Princeton University Press, 1979), 80.

31. Recent experience with AIDS and "emerging diseases," I would argue, has only modified, not invalidated, this generalization.

32. F. Macfarlane Burnet, *Natural History of Infectious Disease,* 2d ed. (Cambridge: Cambridge University Press, 1953), 24. Burnet conceives of infectious disease as "a conflict between man and his parasites which, in a constant environment, would tend to result in a virtual equilibrium, a climax state, in which both species would survive indefinitely" (24).

33. For a synthesis of studies in this area, see Edwin J. Greenlee, "Biomedicine and Ideology: A Social History of the Conceptualization and Treatment of Essential Hypertension in the United States" (Ph.D. dissertation, Temple University, 1989).

34. Steven N. Austad, letter to the editor, *New York Times,* October 3, 1997.

35. Kenneth F. Kiple, *Plague, Pox, and Pestilence* (New York: Barnes & Noble, 1997), 6.

CHAPTER 6

THE NEW ENCHANTMENT

Genetics, Medicine, and Society

SOME YEARS AGO I WAS SPEAKING WITH AN INFLUENTIAL medical chief executive who had moved aggressively to expand his health system's network of physicians and hospitals—while at the same time committing substantial resources and prestige to a gene therapy research program. His response to my question about the risk in such a commitment was a model of policy logic. Only such fundamental research as that exemplified in gene therapy, he explained, could provide a definitive solution to the escalating economic and human costs of chronic illness. Even the most astute market strategies could, at best, only buy time, he argued; the past century's proliferating half-way technologies had simply intensified the problem of ever-expanding costs. Diabetes could be managed, not cured, for example, and the number of diabetics had increased steadily since the introduction of insulin, despite the creation of ever-more-sophisticated techniques for monitoring and treating individual patients.[1] Similarly, in the case of end-stage renal disease dialysis extended life but at substantial economic and often emotional cost. On the other hand, he contended, knowledge of the genetic basis for health and disease could revolutionize the practice of medicine. Gene therapy would provide categorical, not half-way, solutions for chronic ills such as diabetes. Today one finds similarly optimistic arguments justifying investment in stem-cell research; it promises to cure, not simply manage, debilitating chronic disease.

This seemed to me at the time a sanguine, if not utopian, view. It was a vision of sickness definitively vanquished, a vision that promised to dissolve the traditional oppositions that had defined medicine in previous eras: art as opposed to science, mind versus body, idiosyncrasy as opposed to general pattern, reductionist mechanism as opposed to holistic inclusiveness. With enough science, there would be no need for art; idiosyncrasy could be specified and managed in precise molecular terms. The subjective—if educated—judgments of the traditional clinician would be replaced by scientific understanding, by prediction and control. The randomness of sickness would be minimized, if not banished.

This blueprint for the future offers economic and administrative as well as humane rewards. The physician's need to determine and justify individual decisions in terms of aggregate data and institutional constraints (in the form of management guidelines and disease protocols, for example) would be ended. Sufficient basic science knowledge would, in time, dictate every aspect of clinical decision making. All disease would be understood in molecular terms. Skeptics already joke about a future in which every man and woman would have an individual genetic barcode mandating treatment choice.[2] Almost unspoken—yet immanent in this cultural techno fantasy—is the seductive goal of banishing pain and perhaps even death, of reducing emotions and behavior to the malleable working out of discrete neurochemical and neuroanatomic mechanisms.

Whether we regard this vision as utopian or dystopian, it is culturally pervasive. Educated men and women, journalists, even supposedly shrewd venture capitalists and corporate strategists have been enchanted (or have feigned enchantment) by the promise that contemporary genetic medicine seems to offer—a dream that suffuses the ultimately imperfect body with a promise of healing. We live in a moment of what might be called a collective euphoria of magic rationalism, a new style of enchantment with nature. Within the body's frustrating opacity lies a decipherable code, a potential key to life's innermost secrets—a key ultimately accessible to science. It is a vision that fits seductively in a world of chronic disease and an aging population.

Medical progress seems both desirable and inevitable, yet much of the therapeutic payoff implied by basic science findings remains unfulfilled.

Our most significant chronic ills, measured in terms of incidence and mortality—cardiovascular and renal disease, cancer—have thus far resisted simple mechanism-based explanation. In the determination of behavior the tools and concepts of contemporary genetics have similarly proven inadequate to explain and dictate effective responses to emotional pain and socially defined deviance.

Several points, however, seem obvious. One is that the continued search for understanding at the molecular level will prove complex, elusive, and unpredictable. Another is that, once attained, such understandings will pose new social and institutional problems or exacerbate old ones as well as create novel possibilities; it is no more than a cliché to underline the recurring policy challenge of technical change and the need for anticipating—often unanticipatable—consequences. Certainly we are aware of the social, economic, and moral dilemmas created and recreated by technical innovation. Congress, for example, added a formal budgetary line to its genome legislation as a gesture in the direction of studying that research program's ethical, legal, and social consequences.[3] Hundreds of scholars devote themselves to this and parallel problems, and well-funded conferences and panels on aspects of policy, ethics, and technology sprout like mushrooms after weeks of rain. Such widespread attention paid to the dilemmas posed by a novel biotechnology has nevertheless seemed, on balance, marginal when weighed against the laboratory's overwhelming promise.

Knowledge of particular pathological mechanisms, however, does not translate simply and directly into clinical and social practice. Technology alone does not determine clinical outcomes.[4] Every genetic disease has a social history as well as a biological mechanism. I need only point by way of example to the contrasting histories of two such ills: sickle-cell anemia and Tay-Sachs disease. The twentieth-century story of sickle-cell anemia illustrates nicely the interactive and mutually constitutive nature of the social, the technical, and the biological. In sickle-cell the microscope and electrophoresis, as well as the civil rights movement, government lobbying, and attitudes toward race, pain management, and substance abuse were all relevant factors in a complex aggregate that helped constitute the ailment's social history (and thus the experience of individual sufferers).[5] The fortunes of Tay-Sachs in the past generation illuminate a rather different configuration of relevant variables but underline the same pattern of mutual interdependence between the biological, the technical, and the social. In the case of Tay-Sachs, institutions and practices in the orthodox Ashkenazi Jewish community have played a significant role in the context of an increased technical understanding, which created the possibility of screening for carriers and thus community-sanctioned counseling. Both ailments also have a longer—pre-twentieth-century—history, involving the action of ecological and cultural factors over time: in the case of sickle-cell anemia, the macrohistory of selection for malaria resistance; and in the case of Tay-Sachs and certain other "Jewish diseases," the social history of an excluded and inbred community.[6]

Genetic diseases not only differ biologically and in terms of their social profile and visibility, but—like any other ailment—they represent a challenge to individuals as well as to scientists, clinicians, and social policymakers. As we have seen in the past generation's discussions over genetic counseling, communication between physician and patient—or researcher and counselor—is not always easy or unambiguous. The conclusions to be drawn from what is often statistical truth are subject to individual interpretation and idiosyncratic perception. Likelihoods do not dictate decisions. What seems like a grave risk to one individual and his or her family might be ignored or evaded by another. Some men and women are anxious to anticipate their fate; others would prefer to remain in ignorance. The ability to screen for Huntington's or breast cancer genes does not necessarily dictate social or individual decisions—even though it may create the occasion for such decisions. Ominous indications from fetal screening may dictate abortion for some families and individuals, but present a moral and emotional crisis for others, whatever the outcome of their deliberations.

HISTORY VISIBLE AND HISTORY FORGOTTEN

Historians of biology and medicine have in the past generation paid surprisingly little attention to pre-twentieth-century ideas concerning "hereditary disease." Earlier notions of constitution and predisposition appear to contemporary physicians largely irrelevant, unrelated to the developmental trajectory of "valid" knowledge concerning genetics and its relationship to health, disease, and behavior.

This conventional history of intellectual progress has little place for past clinicians' observations. Descriptions of such conditions as hemophilia, polydactyly, and colorblindness seem no more than an interesting set of anomalies and curiosities before they could be construed in post-Mendelian terms. In this sense, the pre-twentieth-century history of ideas about the clinical relevance of heredity is essentially irrelevant to the mainstream narrative of cumulative genetic understanding. In this narrative, genetics—and thus medical genetics—begins with a pantheon of canonical innovators and their achievements: Gregor Mendel and August Weismann; T. H. Morgan's "fly group" at Columbia and its work on "the mechanism of Mendelian heredity"; A. B. Garrod and Linus Pauling on the notion of molecular disease; R. A. Fisher and Sewall Wright on population genetics—and the progressive revelations of what might be called the DNA era. It is a narrative that moves steadily inward from the organism to the cell, from intracellular to molecular mechanisms, ultimately to the association of particular clinical syndromes with such mechanisms. In the conventional narrative, this development of a "properly scientific" understanding of human heredity was sidetracked by the false hopes and misleading goals of eugenics and the ruthless zealotry of National Socialism—events that helped turn academic medicine and biology away from the study of human heredity for two generations.[7] Nevertheless, we can discern a recognizable if fledgling field of medical genetics by the 1960s; a recent history of human genetic disease focuses, in fact, on the period between the mid-1950s and the 1970s as the formative years in the establishment of contemporary clinical genetics.[8] Cytologists and molecular biologists, not practicing clinicians, seem in retrospect the real progenitors of an effective understanding of human heredity and its pathologies.

This is a story of intellectual accomplishment that is too orderly, though, in some sense, not inaccurate. But it is not the only story. It is a narrative largely unconcerned with the day-to-day practice of medicine through the two millennia preceding the First World War.[9] And, perhaps most important from the perspective of understanding both past and present ideas concerning heredity, it is a narrative that legitimates a narrowly intellectualistic and decontextualized understanding of medicine's clinical responsibilities—of the interaction between practitioners and their patients. There is no bedside or consulting room in this history.

Since classical antiquity, physicians have sought to integrate ideas about heredity into their explanations of health and disease and into their rationales for regimen and therapeutics. (In many eras the two could not in fact be easily differentiated). Constitution and predisposition were always important conceptual tools for physicians as they sought to explain and prescribe.[10] Physicians needed to rationalize, to provide hypothetical causes for the ills that afflicted—or threatened to afflict—their patients. In this sense, notions about heredity were an indispensable tool in every nineteenth- and early-twentieth-century practitioner's conceptual armory. They helped explain differential susceptibility to infection and the often unpredictable course of chronic disease; they helped provide a basis for advice on regimen and a rationale for therapeutics in the form of an explanation for idiosyncratic response. Why, for example, did individuals respond differently to the same drug? And why did the clinical course of the same ailment vary so greatly among the men and women who suffered from it? In either instance, constitutional uniqueness could help explain what might otherwise have been awkward inconsistencies in the physician's therapeutic practice and prognostic accuracy.

Although there were comparatively few nineteenth-century monographs on hereditary disease as such, the role of constitution was omnipresent both in the medical literature and among laypeople. "The hereditary transmission of peculiarities of form, intellectual character, manner, and proclivity to disease, is," a mid-century medical authority summarized the conventional wisdom, "no longer a subject of doubt by those best qualified to judge in the matter."[11] Intelligence, temperament, character, all reflected in part an original constitutional makeup—as did predisposition to disease. The traditional rationale for such assumptions seems at first alien to the twenty-first-century reader, and many popular notions of generation seem more like superstition than science. One thinks of the widespread belief in what were called "maternal impressions," the idea that stimuli during gestation could "mark" the fetus permanently and in a manner reflecting the character of that stimulus. Following this thinking, a pregnant woman frightened by a snake might bear a child with a snake-shaped birthmark.

Even the most learned of medical commentators did not and could not distinguish in a categorical way between the influence of heredity and environment. Predisposition to disease, like temperament and idiosyncrasy was the outcome of a cumulative process beginning with conception and subject to modification during gestation, nursing, and early childhood. The body was labile and subject to change as it interacted continuously with its environment; even adult bodies were not entirely fixed, and those cumulative changes would shape the father's or mother's contribution to a new life at the moment of conception. Heredity was a process, not a moment defined by a discrete mechanism—a process with a contingent, but not random outcome.

Diseases seemed traditionally to fall into three categories. Acute infectious diseases, such as smallpox and influenza, were not ordinarily considered hereditary. At the other end of the clinical spectrum, physicians had by the beginning of the nineteenth century, recognized a few seemingly discrete and absolutely hereditary conditions such as colorblindness, but these attracted little attention. No one *doubted* that disease and deformity could be hereditary, so these were seen as atypically categorical, but uncommon and thus clinically marginal, examples of a general and unquestioned truth, too rare and incurable to serve as the conceptual basis for the ordinary physician's thinking about his everyday tasks.

Late-eighteenth and nineteenth-century attitudes toward heredity thus disproportionately reflected the existence of a third class of ailment, the chronic and seemingly constitutional: cancer, gout, mental illness, tuberculosis, and dropsies.[12] Chronic disease lay at the core of the practitioner's concern with heredity. These omnipresent ills served in a practical way to underwrite the general views of heredity I have tried to outline: they seemed inherited, not absolutely, but as gradients or organized tendencies reflecting underlying patterns of constitutional weakness. By the end of the eighteenth century, such chronic ills were often seen as having a substantial hereditary component. The emphasis was on aggregate outcome. One inherited a nervous or scrofular diathesis, but not full-blown madness or tuberculosis; the events of one's life determined whether one would, in fact, succumb. Thus, a physician could remind the son of a gouty father that he must limit his red wine and beef consumption, or he could explain the way in which the stress of a life crisis could precipitate madness in those born with a nervous weakness—while a differently constituted individual could experience the same loss of a loved one or economic disaster yet return to emotional stability after an appropriate period of pain and mourning. An experienced clinician could always link a particular sickness of body or mind with a particular individual's constitutional endowment, social circumstances, and unique personal history. Chronic disease could always be explained—and perhaps even averted through prudent life choices.[13]

Let me illustrate this style of explanation by simply referring to the abundant medical literature on tuberculosis in the late nineteenth and early twentieth centuries. The origin of tuberculosis was most frequently seen as, in part, constitutional yet reflective of environmental risks as well—diet, work conditions, personal habits, family circumstances, stress, ventilation, and housing density could all interact over time to produce particular clinical outcomes. It is hardly surprising that the first quarter century's reactions to Koch's announcement of his discovery of the tuberculosis bacillus in 1882 reflected and incorporated all these long-understood explanatory mechanisms. Constitution continued to play a role in explaining differential susceptibility to Koch's pathogen—interacting with environmental factors such as diet, ventilation, and dirt. The metaphor of seed and soil, much used at the time, incorporated and mirrored the usefulness of the multicausal, process-oriented model I have sought to describe. It incorporated a great deal of empirical observation—both at the bedside and in analyses of vital statistics—in a way that helped explain the undoubted reality of widespread urban exposure to the tuberculosis bacillus, coupled with highly diverse individual outcomes. (The French clinicians' use of the term *terrain* served a parallel explanatory function.) Even in the same tenement house, unrelated families and even family members within the same family might have very different experiences with the disease.

A few relevant details were added to the medical profession's knowledge of hereditary disease in the nineteenth century. A number of entities were identified and shown to be hereditary in a predictable way: Huntington's chorea, for example, or Friedrich's ataxia. But, as in the case of color blindness, the rarity of such absolutely hereditary conditions militated against their having any major impact among practitioners.

The differences between medicine then and now cannot be limited to our far more circumstantial understanding of the body's mechanisms and to our possession of far more powerful tools for probing and imaging that body in health and disease. Equally important, pre-twentieth-century practice was not tied—as it increasingly is today—to bureaucratic institutions and procedures. (An exception might be the way in which nineteenth-century insurance companies urged examining physicians to take note of hereditary tendencies and family histories in deciding whether to authorize policies on particular lives.) Constitutional assumptions helped create an explanatory framework to structure physician-patient interactions; they did not dictate uniform therapeutic choices or reimbursement patterns, mandate referrals and diagnostic procedures, or demand the formulation of patient-sensitive ethical guidelines and governmental policy.

AN OPPOSITIONAL HISTORY

As I have suggested, these hereditarian ideas will seem quite alien to today's practitioners and certainly to medical geneticists. They seem entirely speculative, perhaps functional in past clinical settings, but cultural lubricants, not scientific data. The ideas I have been describing would be dismissed by most physicians today as, in a sense, prescientific. Heredity and constitution were then; genetics is now.

We can, however, cite an alternative historical narrative as well, one in which contemporary genetic medicine is cast in a more ambiguous and less euphoric light. It is the by now familiar narrative of social or humanistic medicine. In this history, our knowledge of the human genome and all it promises in therapeutics and disease prevention exemplifies and constitutes an ultimate reductionism, a culmination of trends that have seen the patient objectified and medicine fragmented, commodified, and bureaucratized—a depersonalized would-be final solution to the problem of subjectivity and biological idiosyncrasy. The physician has become increasingly a node in a grid of laboratory findings, the argument follows, increasingly bereft of agency if not yet of social status. Basic science priorities foreground a handful of rare but scientifically intriguing genetic diseases, whereas less reducible ills—the complex, the multifactorial, the chronic, and the ubiquitous remain, managed at great cost but still ubiquitous.[14] The needs of employers, insurers, and would-be rationalizers of clinical decision making are articulated along ever-changing software pathways as both the physician's and patient's autonomy are reduced to laboratory and imaging findings—determining who is to be treated, how they are to be treated, and who is to survive their uterine environment—ultimately perhaps, who is to be hired or fired.

This oppositional narrative is in fact not entirely inconsistent with the celebratory history described earlier in this chapter. But these alternative stories have a very different emotional and policy valence. One is holistic, in a sense romantic, invoking a world we have lost in which physician and patient related in a personal and multidimensional way—the other focuses on a narrative of inexorable progress through greater understanding of nature in which physician and scientist, medicine and biology become increasingly indistinguishable.

We live in the contested space in which these narratives overlap. The great majority of us welcome medicine's increasing power over disease but regret some of the particular social consequences associated with the accumulation and application of such knowledge. Increasing understanding at the molecular level implies complex, as yet ordinarily unpredictable sequences of events and interconnections in the social as well as biological realms. Such interactions do not stop at the body's physical boundary. One need only read a newspaper article to see how genetic hopes and techniques have become a site of recurrent social contestation.

Economic interests and ambitions constitute one such arena—and an increasingly visible one. The ability to modify crops genetically has not brought an end to controversy, for example, but has occasioned a novel and persistent set of policy debates, both intra- and international. Similarly, the approval of a genetically engineered drug can revive the fortunes of the company that has patented it and offer hope to individual sufferers. We have become accustomed to finding reports of genetic innovation in the business section as well as in the science section—and even on the front page of our newspapers.

There is a compelling irony in the way in which our growing understanding of molecular mechanisms underlines the need for placing that increasingly intimate understanding of the body's molecular fabric in the largest and most interactive of social and political contexts. In treating or advising an individual patient, a physician or genetic counselor must be aware that the set of choices that face both physician and patient are determined not only by our understanding of particular protein sequences but also by factors as diverse as patent law; Washington lobbying; disease-oriented websites; specialty organization and training; drug-company strategies and tactics; conflict-of-interest rules; class, race, age, gender, and individual family variables; and emotionally resonant cultural constraints. It is not surprising that fundamental social conflicts over values and policy should be reflected in the negotiations surrounding genetic research and its applications. I refer to such issues as the appropriate relationship between the public and private sectors, between the market and government as decision maker, and between medicine's traditional notion that it was and is a humane profession and the reality that it was and is a market actor.

ONLY CONNECT: LESSONS FROM AN OBSCURE HISTORY

How can traditional notions of heredity and constitution help us think about such issues? I think they can, and not because they are correct in detail. We cannot restore a lost social and intellectual world, but we can learn some useful lessons by looking at these seemingly antique notions of the body and at the web of social relationships in which constitutional ideas functioned and were themselves reproduced.

This history of constitutional ideas points to some fundamental areas of continued concern. I would like to suggest three, two of which relate to the treatment and explanation of disease, the third to the cultural uses of hereditarian determinism. First is the importance of communication in the day-to-day work of clinical medicine. Ideas about the nature of disease and its management—a shared medical vernacular—have always been central to physician-patient communication. Second is a lesson implicit in the form and content of these seemingly murky traditional ideas. They were built upon the fundamental notion of process and the inclusiveness, linkages, and contingency that the idea of process subsumed. This model provides, in particular, a useful way of thinking about chronic disease, of reminding us of the need to think in terms of interconnections, of a particular outcome reflecting a multiplicity of variables interacting over time. Third is the framework hereditarian ideas have for a century and a half provided for thinking about the nature of human nature and human diversity—explaining phenomena ranging from health disparities to the sources of deviance and diversity. We are what our inherited bodies allow us to be. The explanatory power and cultural prestige of modern genetics has only reinvigorated this tendency toward the cultural invoking of a somatic determinism.

There is something almost irresistible about the power, cultural prestige, and seeming certainty of genetic ideas, whether they promise to explain differential susceptibility to disease or the differences between men and women. The history of older flirtations with eugenic determinisms should make us cautious in judging extrapolations from reductionist mechanism (real or hypothesized, individual or evolutionary) to social or clinical outcome, but we have not escaped this ideological briar patch and have been subjected in recent years to a variety of would-be genetic explanations of highly sensitive behavioral issues. A hypothetical genetic explanation of homosexuality is one example; gender differences in temperament and cognitive abilities are others. Recent genetic explanations of obesity and even binge eating provide still further instances of what might be called a rush to premature determinist judgment. We can expect a regular crop of such reports and an awareness of the long and persistent history of such seductive correlations should make us cautious in evaluating them.

As I have already suggested, constitutional ideas were highly functional in the context of pre-twentieth-century physician-patient relationships. That role suggests the continued need for communication in the microsocial system of healer and sufferer and the problematic quality of that communication in a twenty-first-century world in which lay and medical understandings have diverged dramatically but in which the need for communication remains. Knowledge is not distributed equally, and we need to be very much aware of the human (and sometimes legal and moral) problems built into the management of interactions between physician and patient. It is easy to forget how fundamentally ideas about the body inform such necessary relationships even today: patients always need to make meaningful sense of their situation. The history of medicine makes this lesson unavoidable, and the prognostic potential of modern genetics only raises the stakes in this already difficult but unavoidable relationship. Anyone who has studied the history and practice of informed consent is well aware of such issues; the patient will always constitute a unique individual, neither reducible to probabilities nor humanely engaged and enlightened by the formalized language of formalized consent.

Traditional notions of heredity exemplify more than the need to think seriously about communication between physician and patient (and family). They also provide a way of thinking about the body and society. As we have seen, pre-twentieth-century notions of heredity were built upon the assumption of process, of multiple factors interacting over time—with health, disease, or emotional proclivity being an aggregate resultant of interactions beginning with conception (and before it, in fact, reflecting each parent's life events and choices) through gestation and nursing. We may no longer believe in maternal impressions as understood in the eighteenth century, but we remain concerned with fetal development and the prospective mother's social circumstances. Another aspect of these ideas was the contingency and interconnectedness implied in an emphasis on process; outcomes were not foreordained in a monolithic way but negotiated as particular bodies moved through time from fertilized egg to adult. Physical and psychological environment, diet and stress, all helped shape an individual's internal environment. Body acted on mind, mind on body, the physical and social environment on both. This way of thinking will remain central in an era focused on chronic diseases and on the environmental variables that might interact to produce such ills. It was recently reported, for example, that genetics and stress might be linked to chronic fatigue syndrome.[15] Analyses of asthma and obesity similarly—and by now routinely—link social, genetic, ecological, and policy variables in their discussions of etiology.

I have described two histories of heredity in medicine—one reflecting intellectual developments along a canonical trajectory of knowledge acquisition approaching ever closer to a material substrate in the natural world, the other more clinical, subjective, sensitive to social variables—and increasingly marginal in comparison to mainstream biomedicine. Yet there is something intensely relevant in their emphasis on the multidimensional nature of disease causation. Contemporary history makes unavoidably clear the linkage between genetic knowledge and technology and the social world in which that knowledge is elaborated, diffused, and applied. Lobbyists and patent attorneys as well as bench scientists help determine the precise shape of the therapeutic and explanatory resources available to the individual practitioner. The highly visible ethical issues that recurrently impinge on our collective attention only hint at the need to confront fundamental social, institutional, and even political connections. The recent stem-cell controversy in the United States exemplifies such complex webs of interaction.

Such linkages are only illustrative of the many that cumulatively shape the ultimate social impact of genetic knowledge. The moral and social stakes are as obvious as the demographic and economic. Nevertheless, we have contrived in the past generation to assign many of these questions and relationships to specialists, the bioethicist and the policymaker. But bioethics and policy without historical, social, economic, and cultural context are incomplete and ambiguous. The social implications and consequences of genetic research and its application cannot simply be referred to an ethicist or ethics committee. Nor can we assign them to the policy-oriented health economist, whose particular training provides another variety of often inadequately contextualized analysis.

I would argue that medical men and women must play an active—though obviously not exclusive—role in these collective deliberations, for they are perhaps best suited to track the impact of technical and administrative innovation on the quality of medicine at the bedside, on the impact of the new genetics on individual patients. This is not merely a pious restatement of the traditional argument for the clinician's moral and cognitive responsibility for his or her patients; it is also an analytic position, for that grain of sand that constitutes each clinical interaction also constitutes a key to the universe of ideas and social relationships that ultimately shape that interaction. The twenty-first-century practitioner will need training in social as well as clinical indeterminacy.[16] Modern understandings of genetic contributions to disease (and thus medicine and therapeutics) necessarily dissolve the notion of rigid boundaries between bedside and laboratory, between practice and politics, between the body and the social and material worlds in which it lives and moves.

Process, linkage, interconnectedness, contingency—the characteristic elements in a very old way of conceptualizing the body in health and disease constitute a useful model for thinking not only about the relationship between biopathological mechanism and disease outcome but about the relationship between medicine and society more generally. Clinicians, like the rest of us, need to learn to live with irony—and in particular the paradox implicit in the fact that the most sophisticated advances in knowledge only create new options; they do not end the need for choice.

You will have gathered that I was not entirely convinced by the logic of the medical executive with whose activist hopes for gene therapy I began this discussion. I do not seek romantic solace in the notion of restoring a humane face-to-face doctoring that—notionally—existed in the nineteenth and early twentieth centuries, but I also do not think we will solve the problem of human suffering when in the fullness of time all disease becomes a problem in molecular biology. All of which underlines the continuing centrality of that tradition in medicine that focuses on the physician as social analyst and social actor—and not simply a manager of bodily malfunctions. Genetics and the molecular understandings it is based on promise not an end to social medicine, but a powerful de facto argument for its nurture, evolution, and continued adaptation.

NOTES

1. The much-discussed increase in type-2 diabetes represents another set of etiological—and social-policy—questions.

2. "But when genomes can be decoded for $1,000, a baby may arrive home like a new computer, with its complete genetic operating instructions on a DVD." Nicholas Wade, "The Quest for the $1,000 Human Genome," *New York Times,* July 18, 2006.

3. For a description of the current federal and state policies, see http://www.genome.gov/10001754 (accessed December 2, 2006).

4. Whereas technology, as students of its history have become aware, does not evolve independently of context.

5. See Keith Wailoo, *Drawing Blood: Technology and Disease Identity in Twentieth-Century America* (Baltimore: Johns Hopkins University Press, 1997) and *Dying in the City of the Blues: Sickle Cell Anemia and the Politics of Race and Health* (Chapel Hill: University of North Carolina Press, 2001).

6. Keith Wailoo and Stephen Pemberton, *The Troubled Dream of Genetic Medicine: Ethnicity and Innovation in Tay-Sachs, Cystic Fibrosis, and Sickle Cell Disease* (Baltimore: Johns Hopkins University Press, 2006). See also the parallel case of familial dysautonomia, another "Jewish disease." The historian Susan Lindee describes it "as an experienced and lived disease, is a product of neuronal anomaly, of sensory dysfunction, of Jewish history, and of biomedical technology" *(Moments of Truth in Genetic Medicine* [Baltimore: Johns Hopkins University Press, 2005], 157). Breast cancer research, screening, and treatment provide another highly visible, yet somewhat asymmetrical, instance: attitudes toward gender, toward sexuality, and toward the women's movement played a parallel role in helping determine the changing social shape of this very different disease in which the role of genetic factors is real but difficult to specify.

7. Not all the impediments were political and ideological. Human beings are far from ideal experimental organisms—although the number of men and women suffering from observed clinical syndromes provides a heuristically valuable pool of data ordinarily unavailable in breeding populations less self-conscious and less intensely observed.

8. Lindee, *Moments of Truth in Genetic Medicine*, 2–3. Constitutional medicine survived in a variety of forms in the interwar years but remained, I would contend, an often respectable but minority position. See also, Sarah W. Tracy, "George Draper and American Constitutional Medicine, 1916–1946: Reinventing the Sick Man," *Bulletin of the History of Medicine* 66 (1992): 53–89; Tracy, "An Evolving Science of Man: The Transformation and Demise of American Constitutional Medicine, 1920–1950," in *Greater than the Parts: Holism in Biomedicine, 1920–1950,* ed. Christopher Lawrence and George Weisz (New York: Oxford University Press, 1998), 161–88.

9. Garrod's concept of "inborn errors of metabolism" is something of an exception, growing as it did out of an original clinical perspective. Archibald E. Garrod, *Inborn Errors of Metabolism* (Oxford: Academic Press, 1909; 2nd ed., 1922). The conventional narrative also tends to place comparatively little emphasis on the work of animal and plant breeders—other than in the context of their influence on Darwin and Mendel.

10. Charles E. Rosenberg, "The Bitter Fruit: Heredity, Disease and Social Thought in Nineteenth-Century America," *Perspectives in American History* 8 (1974): 189–235; John Waller, "'The Illusion of an Explanation': The Concept of Hereditary Disease, 1770–1870," *Journal of the History of Medicine* 57 (2002): 410–48, and "Ideas of Heredity, Reproduction and Eugenics in Britain, 1800–1875," *Studies in the History and Philosophy of Biology & Biomedical Science* 32 (2001): 457–89; E. H. Ackerknecht, "Diathesis: The Word and the Concept in Medical History," *Bulletin of the History of Medicine* 56 (1982): 317–25; Elizabeth Lomax, "Hereditary or Acquired Disease? Early Nineteenth-Century Debates on the Cause of Infantile Scrofula and Tuberculosis," *Journal of the History of Medicine and Allied Diseases* 32 (1977): 356–74; Christopher Hamlin, "Predisposing Causes and Public Health in Early Nineteenth-Century Medical Thought," *Social History of Medicine* 5 (1992): 43–70; Carlos Lopez-Beltran, "Human Heredity 1750–1870; The Construction of a Domain" (unpublished Ph.D. diss., King's College, University of London, 1992); J. Andrew Mendelsohn, "Medicine and the Making of Bodily Inequality in Twentieth-Century Europe," in *Heredity and Infection: The History of Disease Transmission,* ed. J.-P. Gaudilliere and Ilana Lowy (London: Routledge, 2001).

11. William A. Hammond, *A Treatise on Hygiene with Special Reference to the Military Service* (Philadelphia: J. B. Lippincott, 1863), 116.

12. Dropsies was the name given to conditions marked by edema; from a modern perspective these accumulations of fluid might most frequently be caused by cardiovascular ills, chronic kidney disease, or possibly tumors.

13. Such explanatory frameworks would have worked particularly well in a "family practice" in which a physician would know family members of different generations over time.

14. As I revised this chapter, for example, my morning paper reported a "High-priced Genzyme Drug is OK'd. 1st Remedy for Rare Pompe Disease," *Boston Globe,* April 29, 2006. The disease affects roughly 1,000 people, the reporter noted, and the cost of treatment is estimated at $200,000 per patient annually. Meanwhile, a variety of mundane but deadly ills in the developing world, as critics repetitively charge, are ignored in a research agenda driven by market strategies shaped by the developed world's disease ecology.

15. "Genetics and Stress Are Found Linked to Fatigue Disorder," *New York Times,* April 21, 2006. The Associated Press report described chronic fatigue as a "condition so hard to diagnose and so poorly understood that some people question whether it is even real."

16. And the factors structuring ultimate outcomes in—for example—cardiovascular disease or asthma are even more complex and difficult to weigh.

-- --

CHAPTER 7

ALTERNATIVE TO WHAT?
COMPLEMENTARY TO WHOM?

On the Scientific Project in Medicine

FOR MORE THAN A DECADE, OBSERVERS OF American health policy have tracked an ongoing conflict over a very small portion of the National Institutes of Health's substantial budget—that which supports the Center for Complementary and Alternative Medicine. I was fascinated in the mid-1990s by the controversy surrounding the origins of this small and generally unobtrusive unit, attracted as much by the vehemence as by the substance of the debate, by the disproportion between heat and light.[1] A representative *New York Times* op-ed piece in 1996, for example, was entitled "Buying Snake Oil with Tax Dollars." The authors—a physicist and a biologist—charged that alternative medicine was no more than a relic of medicine's prescientific past. "It goes by different names now—biofield therapeutics, mental healing, homeopathy—but magic has been used to treat the sick throughout history. The shaman shakes his rattles, the faith healer lays on hands—sometimes patients die, some make surprising recoveries. The only thing that has changed is that the 'healers' of today apparently have the endorsement of the National Institutes of Health."[2] In the summary words of a prominent *Times* science writer, "Some have seen the initiative as visionary, but others liken it to governance by horoscope."[3]

The overwrought tone of the debate has moderated in recent years.[4] Today many American physicians refer patients to acupuncturists, discuss wellness, emphasize nutrition, recommend yoga and meditation.[5] But these examples of therapeutic flexibility obscure an underlying historical continuity, the articulation of categorical boundaries between mainstream and other forms of practice. It is more than a scuffle over profits and who makes them; it is also an argument about the legitimacy and sufficiency of medical authority and—ultimately—the nature of truth. For many physicians and medical scientists, endorsement of traditional herbal remedies, for example, or spiritual healing remains a kind of blasphemy, an affront to the methods, accumulated knowledge, and—not least—the social responsibilities of medicine.[6]

The logic—and title—of my argument begins with a problem of definition. What do we mean by "alternative" or "complementary"? These are not random words. But alternative to *what* is the question. Discussion of "nonstandard" medicine must begin with the standard, the professional community that defines and constitutes what we understand to be legitimate. It is no accident that a widely cited historical survey of nonorthodox medicine is called *Other Healers*.[7] Only one thing unifies such diverse phenomena as homeopathy, Christian Science, acupuncture, prayer, and traditional herbal practices—aside from their being ways of responding to sickness actual or threatened. What they have in common is their otherness, their location outside a social and bureaucratic boundary. Inside that boundary is the enterprise that has come to be called "scientific medicine."[8] Some call it "Western medicine" or "biomedicine," although it is a global and increasingly globalizing system of ideas and practices. And it is a system that incorporates moral value and methodological assumptions as well as seemingly more tangible—and widely admired—laboratory and clinical achievements. This system has also garnered a variety of criticisms within and outside medicine from activists who charge that Western medicine (and its associated commercial and public-sector components) is in thrall to reductionist models, ethnocentrism, and corporate values and has thus defaulted on the profession's traditional moral responsibilities.

It is hard to avoid the use of the word "moral." In the context of this discussion, it is implied by my casual use of the terms "orthodox" and "nonorthodox," with their echoes of distant yet intense theological battles and agendas. A science-based and -validated clinical efficacy has become, for many of us, a self-evident justification for faith in a body of ideas, practices, and professional roles. In this sense, mainstream American medicine is a kind of state church. Those who oppose it have been described since the mid-nineteenth century as members of sects and their interests as sectarian if not simply mercenary. Early-twentieth-century spokesmen for organized medicine often called alternative healers members of cults if they were organized, quacks if they were individuals, superstitious or folk healers if they were associated with particular ethnic groups or localities. In referring to certain practitioners or drug manufacturers as quacks and nostrum vendors, mainstream physicians invoked a demeaning image of duplicity, greed, and exploitation.[9]

Such hostility is understandable. There has been a strain of antiauthoritarianism in much alternative practice, a de facto rejection of credentialed social authority. To practice as a self-consciously oppositional healer has been to question the complex interconnections that link faith with knowledge, and presumed medical efficacy. Truth claims in medicine have never been regarded as abstract, nor are they simply intellectual; they relate not only to social status and technical competence but also to health, even life or death, for ordinary men and women. In the case of therapeutic diversity, more may be less; choices are not, from the mainstream point of view, legitimate options but misguided and—on occasion—even fatal delusions. It is no wonder that debates surrounding alternative medicine have until the present generation been consistently harsh and unrelenting. Any distraction from or denigration of biomedicine can be construed as an antisocial act, a sin against the collective good. Endorsed in the course of the past 150 years by the laboratory's prestige and accomplishments and supported by widespread hopes for future cures, medicine's claims became increasingly absolute during the course of the twentieth century—even as economic competition with noncredentialed competitors became less and less of an issue for the great majority of physicians.

In recent years, however, some cracks have appeared in the ideological and institutional boundaries that separated credentialed practitioners from their competitors. Flexibility and change seem to have become a natural aspect of a complex medical landscape in which everything from the Internet to self-help and advocacy groups, to drug company research and marketing strategies have shaped and are reshaping available therapeutic options and patient expectations. Global population movements have brought new healing practices as well as new women and men to North America. Meanwhile, a quarter century of critical studies has demonstrated an embarrassing persistence of avoidable medical errors and regional practice variations that underline the normal fallibility of mainstream medicine; it can lay no absolute claims to a consistent scientifically validated clinical practice. It might even be argued that a cultural shift toward a more relativistic and antireductionist style of thinking has helped make the walls between establishment medicine and its alternative step-siblings increasingly permeable and therapeutic pluralism increasingly respectable.

Despite such sources and evidences of change, substantial boundaries remain; the power and prestige, the status and accomplishments of scientific medicine cannot be denied. And many physicians and laypeople remain adamant in their disdain for alternative healing practices.[10] On the other hand, support for alternative medicine remains pervasive. There will always be a gap between medicine as a social function—what happens to men and women when they feel ill or fear becoming ill—and the more limited and in some ways arbitrarily bounded enterprise of biomedicine. Alternative medicines will continue to thrive in the ever-shifting space constituted by needs that mainstream medicine cannot or will not—and perhaps should not—consistently address.

MEDICAL TRADITION: IDENTITIES AND BOUNDARIES

In some of its aspects this controversy is relatively modern and associated with the past 150 years. I refer to the rise of what I have called "biomedicine," those linked phenomena associated with laboratories and specialism, with the central role of hospitals, with the cumulative impact of the basic and applied sciences on the clinician's ability to diagnose, explain, and treat disease. The undeniable power of these achievements and institutions seems to underwrite a self-evident distinction between legitimate and illegitimate practitioners, a distinction that has come to seem natural and necessary as well as formal and bureaucratic. Although the elevated status of regular medicine might in historical perspective seem novel and reflect modern attitudes toward science and its accomplishments, efforts to create effective professional boundaries can be traced back twenty-five centuries—to the classical origins of Western medicine.

Let me illustrate this with a reference to what might be the most-cited passage in the history of medicine. It is from a Hippocratic text called "On the Sacred Disease"—or epilepsy.[11] The anonymous author begins by saying that, despite its alarming symptoms, he does not believe "the 'Sacred Disease' is any more divine or sacred than any other disease, but on the contrary, has specific characteristics and a definite cause." The text continues on to explain that the disease, like others, is hereditary: "a phlegmatic child is born of a phlegmatic parent, a bilious child of a bilious parent, a consumptive child of a consumptive parent"—and so in similarly humoral vein. The disease is seated in the brain, the author explains, invoking an elaborate, if speculative, rationalism to explain the ailment's unsettling symptoms. This chronic disease could be cured, the argument follows, as much as other ailments "so long as it has not become inveterate and too powerful for the drugs which are given." The notion that convulsions were evidence of supernatural possession was tainted at its source. "It is my opinion," the author contends, "that those who first called this disease 'sacred' were the sort of people we now call witch-doctors, faith-healers, quacks and charlatans." Hoping to obscure their therapeutic incapacity, such pious pretenders labeled the ailment "sacred" so as to obscure their ignorance and their inability to treat it effectively.[12]

These ancient formulations, though at first glance unrelated to modern understandings, are directly relevant to nineteenth- and twentieth-century notions of the relationship between mainstream medicine and its competitors. First, they illustrate the physician's dependence on a rationalistic and materialistic, textually oriented and systematically transmitted framework for explaining disease—as opposed to explicitly spiritual or nakedly empirical ideas.[13] Second, and perhaps even more fundamentally, they associate that intellectual project with social identity. Priests and charlatans endorse certain ways of thinking about disease causation, rational physicians quite another. There has been a long relationship between the function of healing, the authority of elite practitioners, and the articulation of formal bodies of knowledge embodying and legitimating medical authority.

That legitimacy has been historically related to social status and to learning. Texts and the ability to articulate and deploy them have been keys to the gentleman-physician's identity—along with the social position that facilitated differential access to those texts. Authority grew in part from the mastery of texts, not laboratory or clinical achievements. Thus, the seventeenth-, eighteenth-, even early-nineteenth-century academic physician was more like our professor of English, philosophy, or classics than of biology or physiology; the library was more relevant than the laboratory or postmortem room for the great majority of elite physicians. Equally close was the historical link between the academy and the definition and legitimation of medical status. It is no accident that until the mid-nineteenth century "the faculty" was a synonym for the credentialed medical profession. Nor was it an accident that "empiric" remained a mainstream synonym for quack. Treating patients "empirically," that is without an explicit physiological rationale, was the sign of a medical pretender. Dependence on the authority of individual experience unmediated by a structure of formal ideas was radically egalitarian, an unsettling emphasis on unmediated feeling and intuition over systematic text-based learning. To glorify the authority of experience is to undermine hierarchy—in medicine as in religion. It was no more than logical for popular healers and critical laypeople to ritually disdain the medical man's Latin prescriptions and technical terminology as self-serving mystification.

Until recently—in the West as elsewhere—the healing function has been widely diffused through society. A comparative handful of credentialed medical men could hardly hope to monopolize the market for clinical services, though they did seek to control the high ground of elite paying patients and official and academic appointments (and, after the mid-eighteenth century, hospital attending positions).[14] The world of medical practice before the nineteenth century included bone setters, midwives, herbalists, surgeons, and apothecaries. This was as true of England and the Continent as it was of a provincial North America. Most important, until the past century or century and a half, the bulk of medical care has been provided by laypeople in their homes. Prevailing therapeutic and etiological concepts were multicausal and aggregate, assuming a place for diet, emotions, and regimen as well as drugs. Behavior and biology were also interconnected, for one helped create the other: the mind underlying consciousness, consciousness and conscious decisions helping create a particular body over time through an individual's cumulative pattern of willed decisions. All practice was holistic, and prevention and therapeutics were in some ways indistinguishable.

In this traditional medical environment, credentialed physicians were ordinarily consultants, not primary care givers; they never dominated day-to-day medical care, but defined and defended a (changing) portion of that social activity. Boundary setting was always a priority; learned physicians could not monopolize medical practice, but they could—and did—seek to define and lay claim to portions of it and to rationalize that policing in terms of the mastery of specific bodies of knowledge and adherence to a code of ethical and gentlemanly behavior. Professionalism rested neither on standardized practice nor statistically demonstrable efficacy but turned, instead, on the control of a body of currently plausible natural knowledge and the ability to deploy it in reassuring fashion at the bedside—and, for the most ambitious, to demonstrate that mastery in the written form of books and pamphlets.[15]

In practice this meant incorporating elements from contemporary "science" in the explanatory schemes of educated physicians. I have already referred, for example, to classical humoral theory or—to cite a much later example—the fashion for chemical and Newtonian mechanical models in seventeenth- and early-eighteenth-century medicine. In the early nineteenth century, medicine became increasingly influenced by the prestige and findings of pathological anatomy and the ever-more-pervasive assumption that disease might best be thought of in terms of specific entities with characteristic clinical trajectories and underlying mechanisms.[16] From the mid-nineteenth century on, laboratory findings began to supplement postmortem insights and to promise a more rational, scientifically based medicine. Physiology, physiological chemistry, histology, pharmacology, and—later in the century—bacteriology along with germ theory all worked to change public—and medical—perceptions of what disease was and medicine should be.[17] By the end of the nineteenth century, educated laypeople had assimilated the image of the ideal physician as perhaps not precisely a scientist but as informed in his practice by the laboratory's findings and methods. It was in some ways a novel and status-enhancing image—yet one consistent with an older emphasis on the mastery of esoteric knowledge as essential to the physician's legitimacy. Science promised not only a new level of diagnostic precision but also ultimately a style of practice that was objective, increasingly efficacious therapeutically, and thus necessarily in the patient's best interest.

This resonated with the tradition of the learned physician as gentleman and implied emotional continuity with an even older tradition, that of clerical spirituality, of a distinction between the world of material self-interest and that of selfless professional concern. Of course, medicine in our society is dramatically altered; the home routinely replaced by the clinic and hospital; and medical status and public trust anchored firmly and seemingly necessarily in a bureaucratically credentialed, uniformly trained, and scientifically legitimated profession—not so much in the social identity and presumed moral stature of the practitioner as individual. Ethics has been, in part, outsourced to a specialized subdiscipline of experts.

Nevertheless, I would argue for a significant continuity between the historical figure of the learned and benevolent gentleman-physician and the more recent narrative of physician-scientist as inspired by the quest for objective healing knowledge. This continuity of value has helped provide a compelling rationale for the modern profession's claims to status and autonomy by linking technical mastery with moral responsibility. The laboratory's achievements have underwritten the specialness and legitimacy of a social role that involves touching, seeing, and managing men's and women's bodies by promising that this relationship is in the patients' best interests. Scientific accomplishment has underwritten the physician's clinical autonomy, and American society (the state and states) has granted organized medicine a powerful voice in defining the scope of clinical practice and effective control over membership in its own professional guild—and thus the power to exclude as well as include.

TOWARD A BOUNDED PROFESSION

This hegemony is a novelty in historical terms. Until the end of the nineteenth century, American physicians—like their English and Continental peers—competed for scarce dollars in a crowded and untidy arena. As I have already suggested, primary care was generally provided in every well-ordered home by family members, practitioners of what was often called "domestic medicine," supplemented by the midwives, shopkeepers, and apothecaries who had traditionally coexisted and competed with regular medicine. On slave plantations, West African healing practices survived to compete with the master's and physicians' ministrations.[18] Before the Civil War era, laypeople enjoyed a democratic access to medicine's entire repertoire of drugs and practices—aside from a handful of surgical procedures—as well as the ideas that constituted contemporary professional knowledge. There was little intimidating or inaccessible about managing diets, administering drugs (including those such as mercury, antimony, or arsenic compounds that seem in retrospect alarmingly toxic), or even bleeding.[19] Opium compounds, like mercury-based remedies, were available to anyone who could pay for them. The years before the American Civil War also saw an enormous outpouring of books and pamphlets aimed at individuals and families, providing advice on how to cure and prevent disease, live long, deliver babies, and—sometimes in the same book—avoid their conception. A new generation of health lecturers and reformers promised long and healthy lives to those who would reform their diet, practice the water cure and phrenology, or exercise wisely. Anyone who sold drugs also provided advice about them—and in doing so competed with local physicians, many of whom also sold the drugs they prescribed. In a largely rural society such improvisation and fluidity of roles was only to have been expected.

But in nineteenth-century America a number of alternative healthcare systems emerged, self-consciously challenging the status and epistemological authority of established medicine. Like the sectarian and utopian religious movements that marked American society at that time, they made powerful and categorical appeals to democracy and experience in a world of voluntarism and institutional entrepreneurship. Most prominent among medical "sects" were Thomsonianism and homeopathy. The former was an American creation built around a handful of botanic remedies, and the latter, a German import that thrived in nineteenth-century North America.[20] Hydropathy, or the water cure, a doctrine emphasizing the natural in diet and regimen also thrived in the Anglo-American world of self-management, temperance, and healing spas. What these disparate movements had in common was a willingness to question the safety and efficacy and thus authority of mainstream medicine. They dismissed the medical establishment as ineffective, mercenary, and more mystifying than enlightening in its invocation of learned theory. In their different ways, these energetic reformers drew their inspiration and legitimacy from the authenticity of experience and from the seeming limitations of their "regular" competitors.

American medicine before the Civil War was fragmented and economically marginal. Medical education was sketchy, and the institutional facilities we have come to associate with clinical training were rudimentary. Hospitals and outpatient dispensaries were limited to urban areas in a still overwhelmingly rural society and then served almost exclusively the urban poor. Specialism was still seen as an unethical marketing strategy, and medical education was a mixture of apprenticeship and, at the most, six or seven months of didactic lectures for two years. Treatment was ordinarily provided in the patient's own bed and the physician's technological resources could be carried in his medical bag. The field of practice remained fluid for most of the nineteenth century as self-styled physicians drifted in and out of the profession or practiced part time as they sought to supplement their uncertain medical incomes by farming, keeping shop, or speculating in real estate. Fees were often kept on the physician's books for years—paid in kind or sometimes not settled until a patient's estate was settled. There was indeed a small—disproportionately urban—medical elite who taught in the established schools and monopolized attending positions in the larger hospitals and who had often been trained at least in part in England, France, or, in the last third of the century, Germany and Austria. But such well-educated and economically secure practitioners constituted only a small minority among American physicians in the first three-quarters of the nineteenth century, bound together in part by their mutual respect for the learning and ethical norms that—in theory—they shared.

In the last third of the nineteenth century things changed in a variety of ways. Some new healing systems emerged and survived, among them Christian Science, osteopathy, and chiropracty.[21] A national market for aggressively advertised proprietary drugs was already well established. Physicians sought to inhibit the informal role of pharmacists as primary-care providers, and many sought as well to encourage—and control—nursing, which emerged in this period as a credentialed occupation.[22] In the 1880s, examination-based state licensure began to place a threshold on entry into the profession; before that time there had been little effective medical licensing in the United States.[23] By the First World War a trend had become apparent. The medical profession was becoming increasingly bounded by education and credentials, increasingly in control of the provision of medical care. The number of medical schools and graduates was gradually decreased, and the medical curriculum was upgraded and made more uniform—while the more elaborate training of newly credentialed practitioners legitimated the profession's claims to dominion over medical care. The hospital was becoming central to the provision of that care and the role of regular physicians increasingly hegemonic in these thriving institutions.[24] Nursing was organized and credentialed—but structurally subordinated to physicians. Pharmacists were increasingly constrained in their ability to act as de facto primary-care providers, and drugs were now controlled and divided into prescription-only (so-called ethical) and over-the-counter categories.[25] Some of the older healing systems were being assimilated into the world of mainstream medicine—homeopathy, for example, as its peculiar therapeutic teachings withered and its surviving medical schools (like New York Medical College and Philadelphia's Hahnemann) gradually divested themselves of their founding generations' therapeutic ideology. Osteopathy followed a somewhat parallel course. A willing participant in its own cooptation, it was gradually accepted into the recognized world of medical licensure and practice, though perhaps not into full citizenship.[26] Chiropracty remained in an ambiguous status—still a target of medical anathema into the 1970s, but well established among its lay clientele.[27]

Throughout the twentieth century, the structures of bureaucratic control and credentialing became more elaborate. Specialty boards began to certify practitioners, and hospitals became ever more central to the delivery of medical care—with regular physicians (in practice, board-certified specialists after World War II) controlling access to practice in their wards and rooms.[28] And government, of course, played an increasingly prominent role by the end of the century, approving drugs and subsidizing hospitals, specialty training, health care, and an increasingly significant research establishment. Third-party payers stabilized medical incomes and created another area in which physicians dominated health-care provision.[29] The economics of medicine was no longer precarious as it had been for the majority of practitioners in the nineteenth and early twentieth centuries; competition with irregular practitioners had ceased to be a significant anxiety. The boundaries between credentialed physicians and other healers had become increasingly rigid, yet few questioned the necessity of that demarcation. The firmness of this institutional boundary mirrored the tightness—and thus logically the rightness—of medical identities. The battle over public perception had largely been won.

The prestige of the laboratory and the ideal of scientific medicine dominated the expectations and assumptions of most twentieth-century Americans. Unconventional healers seemed marginal to this world, pious zealots or eccentrics at best, mercenary confidence men at worst. In the world of scientific medicine there are no multiple paths to the truth; and if medicine had not yet discovered the mechanism underlying a particular disease, it ultimately would.

Of course, the twentieth-century social authority of regular medicine was, in fact, never absolute.[30] Families still provided care as did organized health-care reformers (naturopaths, chiropractors, and Christian Scientists, for example), and the management of emotional and behavioral ills remained a contested area with lay analysts, clinical psychologists, psychiatric social workers, and pastoral counselors all playing a role in addition to board-certified psychiatrists. Alcoholics Anonymous provided another kind of nonmedical healing system. Surveys of patient choice in the 1920s and 1930s demonstrated a persistent clientele for irregular practice—a constituency whose tenacity dismayed regular medicine.[31] It seemed wrong that patients should fail to take advantage of the knowledge and skills that science had provided twentieth-century clinicians, especially since medicine could now intervene in the course of disease in ways that previous generations of practitioners could not. Diphtheria anti-toxin, x-rays, insulin, the sulfa drugs, antibiotics, and steroids all reiterated and reaffirmed that compelling truth in the half century between the 1890s and 1940s. Laypeople might not recognize the science underlying biomedicine, but they understood the idea of specific disease as specific nemesis with a characteristic trajectory in which their physicians could—or would in the future be able to—intervene. Reductionist models of disease implied and justified the medical profession's clinical authority.

The American Medical Association and other enemies of irregular practice had since the Progressive era dismissed sectarian competitors as not only mercenary, but also dangerous, since they kept sufferers away from effective care until it was too late. Such arguments were, for example, the bread and butter of the American Society for the Control of Cancer as well as organized medicine. Early detection and surgical intervention was the only way to stop cancer's otherwise relentless course.[32] "Quackery" remained the term of choice for most spokesmen for regular medicine when they assailed alternative healers. Perhaps the most irritating deviants were the organized and well-entrenched sects, most visibly osteopathy and chiropractic. The latter remained in the AMA's opinion an "unscientific cult," in the words of a 1966 pamphlet, "whose practitioners lack the necessary training to diagnose and treat human disease."[33]

THE STATE OF PLAY: MANAGING DIVERSITY

The past generation has indeed seen a softening of mainstream rhetoric and even referral practice. American medicine seems more open to a variety of therapeutic and preventive approaches, ranging from acupuncture and massage therapy to hospice care. But such changes are difficult to evaluate. How flexible and genuinely pluralistic is our health-care system? Do a thousand flowers bloom in the gardens of contemporary medicine? Some years ago, the prominent students of alternative medicine, Ted J. Kaptchuk and David M. Eisenberg, were optimistic about such possibilities. "The dissolving of a single modernist Medical narrative," they wrote in the *Annals of Internal Medicine* in 2001, "has formed an increasing awareness of medical pluralism. The older cultural war of a dominant culture versus heretical rebellion in politics and religion as well as medicine has begun to transform into a recognition of postmodern multiple narratives. . . . Perhaps because it is beleaguered from battles on other fronts, orthodox medicine has simply abandoned its crusade against alternative medicine. . . . A cease-fire, if not a complete armistice, has been declared." It is a dramatic change, they conclude, "away from antagonism and toward a postmodern acknowledgement of diversity."[34]

Perhaps this is true for some of us, but not for most members of the medical community.[35] For most of my contemporaries, the "modernist narrative" remains the only firm—and moral—basis for a responsible profession. Multiple narratives and contingency are the staples of English professors, not board-certified practitioners, yet it is true that the mix of medical attitudes has changed since the 1970s and 1980s, when the great majority of physicians still saw alternative—and complementary—medicine in dismissive terms, condescending at best and more often actively hostile.[36]

The anxieties of clinicians today are increasingly focused elsewhere. The most urgent contemporary threats to the profession are more fundamental and implicate the profession itself as unwitting co-conspirator. Economic constraints and the aggregate truths of randomized clinical trials and clinical epidemiology have eroded the physician's day-to-day clinical autonomy.[37] The individual physician's traditional skill in managing idiosyncrasy seems less and less a necessity—and more and more a form of self-mystification. Physicians often feel powerless in the coils of bureaucratic and corporate structures. Pharmaceutical companies have also come to play a significant role in both developing products and—in tandem—the markets for those products. The power of the market and of the logic of bureaucratic rationalization—the practice guidelines, pressures for quality assurance, and the like—place far greater constraints on contemporary clinical decision making than the competition of alternative healers. Algorithms now seem more threatening than acupuncture or massage therapy.[38]

In the face of such a variety of change and the presence of powerful new stakeholders in the struggle for control of medical decision making, many physicians have, as we have seen, softened their traditionally disdainful attitudes. A partial and uneasy truce prevails between regular medicine and its alternatives, but it is one based on accommodation and convenience, not on an epistemological egalitarianism or relativist pluralism. This ceasefire is structured instead around three related strategies: tolerance, outsourcing, and salvage.

Let me explain my terminology. "Tolerance" implies just that: allowing a variety of "irregular" practices to exist. The word also implies condescension, a willingness to countenance—and even cooperate with—alternative and complementary practitioners, but not on a basis of equality. Professionals can grant such tolerance, but they cannot grant ultimate—epistemological—equality without denying the validity of their own system of knowledge acquisition and application. Tolerance acts out and thus affirms hierarchy. "Outsourcing" is a term I use to describe the need for allies in dealing with the enormous and persistent clinical burden of chronic illness and incapacity—allies who may or may not cure, but who care (and often at a relatively modest cost). Despite its increasing willingness to accept—and even refer patients to—alternative practitioners, regular medicine retains its right of clinical first refusal. Finally, there is the notion of "salvage"—of an ongoing rummage through the world of traditional herbs and medical practices for things that can be identified, validated, and brought into "our world" of randomized clinical trials and measurable efficacy.[39] Does a particular herb "work" when subjected to such critical scrutiny? This was initially the explicit basis for peaceful co-existence between the National Institutes of Health's Center for Complementary and Alternative Medicine and the rest of the NIH and for countless discussions of traditional and alternative medicine since. Stephen Straus, the incoming head of the Center for Complementary and Alternative Medicine in 2000, explained to a reporter from *Science,* for example, that he was "most enthusiastic about clinical trials of compounds, such as botanicals or shark cartilage, because these can be tested in double-blind, placebo-controlled trials—the height of scientific rigor."[40] That which can be seen as efficacious in biomedical terms can be rescued from the tropical rainforest (sometimes literal and always metaphorical) of practice and belief outside the boundaries of scientific validation.[41] There is a long history to be cited here, from quinine and smallpox vaccine to rauwolfia and artemesia.

What these rhetorical strategies have in common is their assumption of hierarchy and their emphasis on maintenance of regular medicine's right to ultimately define the true and the efficacious in the profession's own terms. It seems to me that the boundaries between regular medicine and every other aspect of health care and prevention remain firmly in ideological place—even if a de facto pluralism is acted out every day in a variety of ways and locations. The world of mechanism, of molecular biology and randomized clinical trials, of specific diseases and discrete causes, of medicine as would-be science is our world of value, reference, and expectation. It will not go away.[42]

SOURCES OF PATRONAGE

Nor will its alternative and complementary shadow. A variety of healing and preventive practices will always exist in that space created by the gap between medicine as the broad social function of care and healing and that smaller territory defined and claimed by medicine as a profession. That unmet need guarantees a clientele for informal and irregular modes of care, cure, and prevention. Continued support for such practices has a variety of predictable social, cultural, and demographic sources, and, for the sake of analysis, I will cite seven such factors, with the understanding that they are necessarily linked in complex and contingent ways.

First is disease prevalence itself. In an era of increasing age and chronic disease, there will be a continuing burden of recalcitrant ills, such as some end-stage cancers, that are as yet untreatable, and others, such as diabetes, asthma, and chronic kidney disease, that demand long-term multidimensional care. In addition there are chronic, non-life-threatening conditions that are nonetheless persistent and resistant to treatment. I refer to ills ranging from arthritis, sleep disorders, and depression to irritable bowel, migraine, and chronic back pain. Mainstream physicians often have little time—or sometimes motivation—for dealing with such cases. Our emphasis on technologically sophisticated procedures and drugs allied with a (not unrelated) drive for cost controls guarantees the low priority that most clinicians allot to such ubiquitous chronic ills.[43]

Second is the persistence of underserved populations, a reflection of residential, class, and ethnic diversity. Such differences express themselves in patient expectations and available treatment options. Asian and Hispanic immigrants, for example, may assume particular notions of disease causation, treatment, and prevention—concepts unfamiliar or alien to mainstream American physicians and to American medical education. Recent discussion of what has come to be called "cultural competence" remains a marginal—even decorative—element in American clinical training.[44]

A third factor grows out of the subjectivity of symptoms and the—parallel and paradoxical—rigidity of diagnostic categories. Individuals will always want to own their own symptoms, and for many ailments the skepticism of the medical world toward "borderline ailments," toward such ills as chronic fatigue syndrome or chronic Lyme disease, toward back pain or irritable bowel syndrome impels men and women to seek help—and a sympathetic legitimacy for their ills—from one or another healer with the time or the ideological commitment—and, cynics might add, low overhead—appropriate to treating their ills. Acknowledging and managing chronic pain has never been a consistent strength of the regular profession.

Fourth, mainstream medicine has for well over a century displayed a subtle and sometimes not so subtle bias—a kind of priority gradient—that devalues or assigns a lower status to mind-body relationships and to ailments whose manifestations are primarily emotional or behavioral. Some surveys of alternative medicine use, for example, ingenuously but revealingly include psychotherapy and group therapy among their lists of alternative modalities. Although physicians have always conceded the role played by mind and emotions in shaping sickness and health, such matters remain an area of tension within regular medicine.[45] Even when the purveyors of psychiatric expertise are credentialed physicians—members of the club—many other clinicians have for the past century felt uncomfortable or ambivalent about dealing with ailments manifestly or latently "functional." Psychiatry has been the medical profession's residual legatee of a variety of ills that do not fit neatly into the categories of reductionist medicine—thus acting out the marginal status of psychiatry itself in relation to medicine generally. Contemporary emphasis on psychopharmacological treatments and neurochemical models in the management and understanding of emotional ills expresses as much as resolves this stigmatizing set of priorities.

Fifth is the concern for diet and regimen expressed in a number of alternatives to mainstream practice. Although medicine from classical antiquity had always foregrounded diet and "lifestyle" in explaining disease causation and treatment, modern medicine gradually shifted its focus away from such issues. Only in the past generation have those difficult-to-control etiological and therapeutic factors been restored to medical favor, but they remain elusive and, I would contend, marginal to the practice of most physicians. The emphasis on diet especially has remained and will, I believe, continue to remain central to a variety of lifestyle-reform efforts such as vegetarianism (in its several varieties) and nutritional systems and supplements promising health or physical attractiveness.[46] The appeal of prospectively managing one's own medical fate cannot be expected to diminish, especially in an era of chronic disease.

A sixth source of continued patronage for alternatives to mainstream medicine grows out of what can be called a search for meaning. The consolations of mechanism and randomness are not emotionally and conceptually adequate for some of us all of the time and for most of us some of the time. There are, of course, many sources of alternative frameworks for construing the incidence of health and disease. Some are traditionally spiritual, others secular, but all tend to look beyond the body as isolated biomolecular entity. The persistent desire for meaning beyond mechanism creates a constituency for holistic frameworks, for vitalistic and integrative schemes that link the individual to his or her social and material circumstances and that construe the body's internal environment as a balanced and ever-rebalancing system in a necessarily ongoing relationship with its external environment.

A seventh factor grows out of unresolved and perhaps unresolvable conflicts within medicine itself. How does one weigh idiosyncrasy against general guidelines? Economic rationality against emotional need? Physiological as opposed to social or spiritual efficacy? Physicians as well as patients want incommensurate things: technical capacity and predictable efficacy in a setting that also provides personal attention and thoughtful care—organ transplants and genetic therapy with a warm and caring face (and time to listen).

No foreseeable developments in medicine's technical capacity will resolve these structured contradictions, and the supply of individuals seeking help outside the boundaries of regular medicine is unlikely to diminish.

CONCLUSION: MIRRORS OF MEDICINE

Why am I, as a historian, concerned with this elusive problem? I do not go to alternative healers and have, in fact, a good deal of—appropriately cautious—faith in the laboratory and in the generalized truths of mainstream medicine. I am not a patron of chiropractic (though, as the proprietor of a quirky back, I might become one) or herbal medicine. I am concerned because I remain fascinated by that constantly renegotiated space that I have already described, that which separates perceived health needs—medicine as a social function—from that circumscribed portion of it occupied by mainstream biomedicine. What regular medicine has become determines how large that social space will be and the men and women who will occupy or move through it.

Alternative medicine is, in other words, an index to the social and cultural geography of regular medicine. It illuminates needs unaddressed in our health-care system and aspects of our system not easily reduced to that which can be validated, measured, expressed in data-derived algorithms.[47] Alternative practice reminds us that biomedicine, too, is a culture—historical, negotiated, contingent—and not simply a necessary institutional expression of scientific knowledge and technical capacity. And it is an aspect of this culture's power that its contingency is ordinarily invisible to those who dwell within it. Medicine need not be what it is.

Even within mainstream medicine there are efficacies and modes of action invisible to randomized clinical trials and often relegated to that second-class epistemological and social status of placebo effect or patient management. Alternative medicine makes us think rather differently about regular medicine.[48] Medicine is caring as well as curing, ritual perform-ance as well as scientific understanding and physiologically measurable procedure.[49] "Alternative medicine might," as one of its sympathetic stu-dents put it, "instead be an opportunity for medicine to reflect on how its scientific apparatus needs to be tempered by magic—even if this magic is usually called 'art.'"[50] That traditional healing systems might "work" does not figure prominently in most contemporary discussions; the "efficacy" of such healing systems has been both casually conceded and method-ologically dismissed by Western medicine for centuries.[51] The basis for such effectiveness has been defined away—depending on the physician's era and language—as no more than sympathy, or suggestion, or "only" placebo effect, as though these were mere confounding variables and not significant aspects of a functioning medical system. "Medicine is healing," in the words of medical historian Owsei Temkin, "based on such knowl-edge as is deemed requisite. The fact that medicine in our days is largely based on science does not make other forms less medical—though it may convince us that they are less efficacious."[52]

There will always be a distinction, as I have suggested, between sci-entific credentialed medicine and the rest of care giving, but a corollary is that the boundaries between those sectors will change with changes in technology, social policy, and the relevant basic science. Just as disease is, to a degree, socially constructed and our knowledge of it changing, so is the physician's identity—and the clinical practices that both justify and in part constitute that identity. What are we to make of the Internet as a source of medical advice, disease advocacy groups, the advertisements and physician visits of the drug companies? Is hospice a part of regular medicine—or an example of an institution defining and mediating a bor-der controversy? The boundaries between establishment medicine and its rivals is at once rigid and permeable. These complex interdependencies and interconnections demand our attention and make us realize that a health-care system cannot be reduced to economic relationships, institu-tional structures, and the application of laboratory findings. But such questions are often invisible to mainstream medicine—as though medicine were an array of methods and data and not a culture itself with a particular and far-from-inevitable history.

NOTES

1. The center was originally, and even more modestly, initiated as the Office of Alternative Medicine. For a useful account of the early history of this entity, see James Harvey Young, "The Development of the Office of Alternative Medicine in the National Institutes of Health, 1991–1996," *Bulletin of the History of Medicine* 72 (1998): 279–98.

2. Robert L. Park and Ursula Goodenough, "Buying Snake Oil with Tax Dollars," *New York Times,* January 3, 1996.

3. Natalie Angier, "U.S. Opens the Door Just a Crack to Alternative Forms of Medicine," *New York Times,* January 10, 1993.

4. James Whorton, a prominent student of this history, reported that he had offered an elective course on alternative medicine to University of Washington medical students since 1986. "Initially," he recalled, "the project seemed a bit like teaching druidism in a Christian Sunday school, although my object never was to convert students to unconventional medicine." Whorton saw his historian's role as contributing to a "process of conciliation." From *Nature Cures: The History of Alternative Medicine in America* (Oxford: Oxford University Press, 2002), xi.

5. José A. Pagán and Mark V. Pauly, "Access to Conventional Medical Care and the Use of Complementary and Alternative Medicine," *Health Affairs* 24 (2005): 255–62; David M. Eisenberg, Roger B. Davis, Susan L. Ettner, Scott Appel, Sonja Wilkey, Maria Van Rompay, and Ronald C. Kessler, "Trends in Alternative Medicine Use in the United States, 1990–1997," *JAMA* 280 (1998): 1569–75; John A. Astin, "Why Patients Use Alternative Medicine: Results of a National Study," *JAMA* 279 (1998): 1548–53.

6. This struggle also resonates with the past decade's so-called science wars, a murky yet intense conflict about the nature, the social authority, even the epistemological status of the scientific enterprise.

7. Norman Gevitz, ed., *Other Healers: Unorthodox Medicine in America* (Baltimore: Johns Hopkins University Press, 1988). For more recent historical overviews of the American scene, see Robert D. Johnston, ed., *The Politics of Healing: Histories of Alternative Medicine in Twentieth-century North America* (New York: Routledge, 2004), and Whorton, *Nature Cures.*

8. All of which indicates the difficulty of making and sustaining categorical definitions of *alternative* and *complementary.* The tactic of distinguishing between "alternative" as implying oppositional and "complementary" as implying active collaboration (and subordination) does not clarify the problem.

9. See, for example, this orientation as illustrated in the titles of James Harvey Young's pioneer studies in this area: *The Toadstool Millionaires: A Social History of Patent Medicines in America before Federal Regulation* (Princeton, N.J.: Princeton University Press, 1961); *The Medical Messiahs: A Social History of Health Quackery in Twentieth-Century America* (Princeton, N.J.: Princeton University Press, 1967); *American Health Quackery: Collected Essays* (Princeton, N.J.: Princeton University Press, 1992). Venality was the key theme in this way of understanding irregular healers and healing. "For the physician seeks to help his patient if he can," as Young put it, "but must sometimes confess that he cannot. The quack need make no such confession, because for him integrity is not, as for the good physician, a sine qua non." The quack would "treat people kindly" and "promise them anything, a regimen that has long lured victims from among the uninformed, lonely, disturbed, and desperate." From Young, "The Persistence of Medical Quackery in America," *American Scientist* 60 (1972): 318.

10. As I was editing this chapter, for example, the Sunday supplement of my local newspaper published an essay dismissing alternative medicine in such familiar terms. "We may laugh at the gullible public from a century ago that was taken in by the hucksters of patent medicines, but, in truth, little has changed. We are still remarkably credulous. Thus millions buy homeopathic medicines, go to acupuncturists for anything that bothers them, believe in aromatherapy, take megadoses of vitamins, and routinely undergo chelation therapy. In most cases and for most uses, each of these therapies is largely scientific nonsense whose reputation for effectiveness is a combination of questionable anecdotes and wishful thinking—the well-known placebo effect." From Tom Keane, "Healthy Skepticism," *Boston Globe Magazine,* October 8, 2006, 12.

11. I use *epilepsy* as a generic term to describe what must have been—from a modern clinical perspective—a variety of conditions marked by convulsions. The classic history remains Owsei Temkin, *The Falling Sickness: A History of Epilepsy from the Greeks to the Beginnings of Modern Neurology,* 2nd ed. (Baltimore: Johns Hopkins Press, 1971, orig. pub. 1945).

12. *Hippocratic Writings,* ed. with an introduction by G. E. R. Lloyd, translated [from the Greek] by J. Chadwick and W. N. Mann et al. (Harmondsworth, England: Penguin, 1978), 237–41.

13. The unchanging elements in this tradition, to put it another way, are representation (in words, images, and in more recent years equations) and rationalization, that is the explanation of felt sickness in terms of some body of more general theory.

14. By the term *official,* I refer to positions such as court or city physician.

15. I have argued elsewhere that traditional medicine did rely on a visible physiological—and in this instance social—efficacy, inasmuch as it employed drugs such as emetics and cathartics that had a dramatic, visible, and predictable effect in seeming to bring the body back into a health-constituting physiological balance.

See Rosenberg, "The Therapeutic Revolution: Medicine, Meaning and Social Change in Nineteenth-Century America," *Perspectives in Biology and Medicine* 20 (1977): 485–506.

16. For a summary discussion, see "The Tyranny of Diagnosis: Specific Entities and Individual Experience," chapter 2 in this volume.

17. They also inspired the creation of a rhetorically compelling history celebrating the emergence of a progressively valid understanding of the body and the achievements of science's saints and martyrs—Newton, Harvey, Giordano Bruno, and Galileo. Sectarian spokesmen often invoked the same narrative of truth persecuted or misunderstood to dramatize their group's relationship to an oppressive regular medicine (homeopaths called regulars allopaths, implicitly designating them as just another striving sect).

18. See, for example, Sharla M. Fett, *Working Cures: Healing, Health, and Power on Southern Slave Plantations* (Chapel Hill: University of North Carolina Press, 2002).

19. There is a certain paradox here. The overlap between lay and medical understandings was functional, in the sense of facilitating belief in the physician's ministrations, but it worked at the same time to undermine hierarchy and thus the exclusive claims of the credentialed physician.

20. For studies of Thomsonianism and homeopathy, see Alex Berman and Michael A. Flannery, *America's Botanico-Medical Movements: Vox Populi* (New York: Pharmaceutical Products Press, 2001); John S. Haller, Jr., *The People's Doctors: Samuel Thomson and the American Botanical Movement, 1790–1860* (Carbondale: Southern Illinois University Press, 2000); Haller, *Kindly Medicine: Physio-Medicalism in America, 1836–1911* (Kent, Ohio: Kent State University Press, 1997); Martin Kaufman, *Homeopathy in America: The Rise and Fall of a Medical Heresy* (Baltimore: Johns Hopkins University Press, 1971); Harris L. Coulter, *Divided Legacy: The Conflict between Homeopathy and the American Medical Association: Science and Ethics in American Medicine, 1800–1914* (Richmond, Calif.: North Atlantic Books, 1973); Anne Taylor Kirschmann, *A Vital Force: Women in American Homeopathy* (New Brunswick, N.J.: Rutgers University Press, 2004); Naomi Rogers, *An Alternative Path: The Making and Remaking of Hahnemann Medical College and Hospital of Philadelphia* (New Brunswick, N.J.: Rutgers University Press, 1998); John Harley Warner, "Medical Sectarianism, Therapeutic Conflict, and the Shaping of Orthodox Professional Identity in Antebellum American Medicine," in *Medical Fringe and Medical Orthodoxy 1750-1850*, ed. William F. Bynum and Roy Porter (London: Croom Helm, 1987).

21. One might also add the Seventh-day Adventist denomination in which healing has always played a prominent role. See Ronald L. Numbers, *Prophetess of Health: Ellen G. White and the Origins of Seventh-day Adventist Health Reform*, rev. ed. (Knoxville: University of Tennessee Press, 1992 [1976]); and Rennie B. Schoepflin, *Christian Science on Trial: Religious Healing in America* (Baltimore: Johns Hopkins University Press, 2003).

22. Some physicians opposed the movement for nursing registration as creating a potential rival for credentialed clinical authority.

23. Samuel Baker, "Physician Licensing Laws in the United States, 1865–1915," *Journal of the History of Medicine* 39 (1984): 173–97; and, more generally, Richard H. Shryock, *Medical Licensing in America, 1650–1965* (Baltimore: Johns Hopkins Press, 1967).

24. See Charles E. Rosenberg, *The Care of Strangers: The Rise of America's Hospital System* (New York: Basic Books, 1987), and Rosemary Stevens, *In Sickness and in Wealth* (New York: Basic Books, 1989).

25. The Harrison Act of 1914 controlled the sale and use of narcotic drugs, in itself an important change in medical practice and lay self-management.

26. Osteopathic schools retained their name and continued to teach their characteristic form of practice—along with a more standard curriculum. Norman Gevitz, *The DOs: Osteopathic Medicine in America,* 2nd ed. (Baltimore: Johns Hopkins University Press, 2004).

27. Norman Gevitz, "The Chiropractors and the AMA: Reflections on the History of the Consultation Clause," *Perspectives in Biology and Medicine* 32 (1989): 281–99. James Whorton has pointed out that Congress appropriated $2 million to investigate the "scientific basis" of chiropractic, the same year that it was approved for Medicare reimbursement (*Nature Cures,* 294). See also Steven C. Martin, "'The Only True Scientific Method of Healing': Chiropractic and American Science, 1895–1990," *Isis* 85 (1994): 206–27.

28. Rosemary A. Stevens, "Medical Specialization as American Health Policy: Interweaving Public and Private Roles," in *History and Health Policy in the United States: Putting the Past Back In,* ed. Rosemary A. Stevens, Charles E. Rosenberg, and Lawton R. Burns (New Brunswick, N.J.: Rutgers University Press, 2006), 49–79; Stevens, *American Medicine and the Public Interest,* rev. ed. (Berkeley: University of California Press, 1998 [1971]); George Weisz, *Divide and Conquer: A Comparative History of Medical Specialization* (Oxford: Oxford University Press, 2005).

29. I refer to Blue Cross and Blue Shield beginning in the 1930s and to Medicare and Medicaid after 1965.

30. See, for example, Michael S. Goldstein, "The Persistence and Resurgence of Medical Pluralism," *Journal of Health Politics, Policy, and Law* 29 (2004): 925–45. Dentists, nurse practitioners, and podiatrists also made claims to caring for the sick and preventing illness.

31. For a well-informed contemporary evaluation, see Louis S. Reed, *Midwives, Chiropodists, and Optometrists: Their Place in Medical Care* (Chicago: University of Chicago Press, 1932) and *The Healing Cults: A Study of Sectarian Medical Practice: Its Extent, Causes, and Control* (Chicago: University of Chicago Press, 1932). Reed's studies constituted the fifteenth and sixteenth publications of the Committee on the Costs of Medical Care. Contrast this with Edward J. G. Beardsley, "Why the Public Consult the Pseudo Medical Cults," *Journal of the Medical Society of New Jersey* 21 (1924): 275–81. "The medical cults have always existed in some form or other and probably always will. We must, as a profession, see to it that they do not exist as a protest against our failure to meet the needs, actual and psychical, of the public"

(281). Studies of the extent of support for alternative medicine in the 1990s elicited a parallel surprise, even among experienced clinicians.

32. Robert A. Aronowitz, "Do Not Delay: Breast Cancer and Time, 1900–1970," *Milbank Quarterly* 79 (2001): 355–86.

33. American Medical Association, Department of Investigation, *Chiropractic: The Unscientific Cult* (Chicago: AMA, 1966), 3. "Since the birth of chiropractic in 1895, the medical profession has been warning the public of the hazards involved in entrusting human health care to these cult practitioners." Chiropracty constituted a primary target of the AMA's Committee on Quackery for much of the twentieth century.

34. Ted J. Kaptchuk and David M. Eisenberg, "Varieties of Healing. 1. Medical Pluralism in the United States," *Annals of Internal Medicine,* August 2001, 193. See the parallel remarks by Robert D. Johnston in his introduction to *Politics of Healing,* 2. "No longer can we simply assume a decades-long belief in the epistemological legitimacy of the medical model, to the exclusion of therapeutic alternatives. Nor can we take for granted the marginality of non-orthodox treatments as if their continued existence was for so many years merely a story of vestigial curiosities, oddities to be pulled off the shelf and gawked at much like an exhibit from an early natural history museum."

35. "There is no such thing as 'alternative' medicine," the president of the American Medical Association explained in the *New York Times*. "There is only scientifically proven, evidence-based medicine supported by solid data or unproven medicine, for which scientific evidence does not exist." From J. Howard Hill, letter to the editor, February 11, 2006.

36. For a sample of the varied attitudes in academic medicine, see Mary Hager, ed., *Education of Health Professionals in Complementary/Alternative Medicine* (New York: Josiah Macy Jr. Foundation, 2001).

37. Stefan Timmermans and Marc Berg, *The Gold Standard: The Challenge of Evidence-Based Medicine and Standardization in Health Care* (Philadelphia: Temple University Press, 2003); Marc Berg, *Rationalizing Medical Work: Decision-Support Techniques and Medical Practices* (Cambridge, Mass.: MIT Press, 1997). The same rapidly evolving technology that has increased faith in medicine's therapeutic capacities has itself increased costs and thus—ironically—ultimate pressures for physician-constraining cost controls.

38. Late-twentieth-century criticisms of the arbitrariness of everyday clinical practice and its regional and specialty differerences have been widely acknowledged but have done little to delegitimate the authority of regular medicine; in fact, such internal reform efforts, like the more recent vogue for evidence-based medicine can be construed as in effect defending, legitimating, and defining the boundaries between science-based providers and those other healers outside the bounds of epistemological respectability.

39. One might also compare the non-Western cultures from which particular herbs have been salvaged to the soil in which antibiotics such as streptomycin have been found. They are sources, not actors.

40. Erik Stokstad, "Stephen Straus's Impossible Job," *Science* 288 (June 2, 2000): 1569.

41. Several years ago, for example, the World Health Organization began a study of non-Western healing practice. "Dr. Ossy Kasilo, the agency's specialist in African remedies, said its first task was 'to just undertake an inventory, then do research on what works and what doesn't.'" Donald G. McNeil Jr., "With Folk Medicine on Rise, Health Group Is Monitoring," *New York Times,* May 17, 2002.

42. Even among those utilizing alternative medicine, only a small minority report having relied primarily on such modes of treatment. See, for example, Astin, "Why Patients Use Alternative Medicine," 1548. The study found that only 4.4 percent of patients fell into this category.

43. Paradoxically, of course, the demographic and epidemiological transition that has produced longer life spans and thus a greater burden of chronic and altered disease is to some extent a consequence of the effects of medicine and public health in parallel with technological and economic growth generally. Whatever the debates among historians and demographers as to the appropriate weight to assign these several variables in bringing about change in morbidity and mortality, most laypersons—and physicians—credit the accomplishments of scientific medicine with a key role in creating today's vital realities. Such widely shared expectations of medical technology do not mean that dissatisfied patients will not turn from regular medicine to other practitioners when motivated by continued pain or incapacity.

44. For a widely read example of such concerns, see Anne Fadiman, *The Spirit Catches You and You Fall Down: A Hmong Child, Her American Doctors, and the Collision of Two Cultures* (New York: Farrar, Straus, & Giroux, 1997).

45. "Quackery seems to have here got hold of a truth which legitimate medicine fails to appreciate and use adequately. Assuredly the most successful physician is he who, inspiring the greatest confidence in his remedies, strengthens and exalts the imagination of his patient: . . . Entirely ignorant as we are, and probably ever shall be, of the nature of mind, groping feebly for the laws of its operation, we certainly cannot venture to set bounds to its power over those intimate and insensible molecular movements which are the basis of all our visible bodily functions." From Henry Maudsley, *Body and Mind: An Inquiry into Their Connection and Mutual Influence, Specially in Reference to Mental Disorders* (New York: D. Appleton, 1871), 39–40.

46. Perhaps the most influential of such movements was temperance (which was, after all, a medical as well as moral reform).

47. Critiques by socially or culturally oriented critics have, on balance, remained revealingly oppositional. They have been not so much dismissed as ignored in the practices and priorities of mainstream medicine.

48. It is also an occasion for comparative historical and sociological analysis. Even within the West, national histories have created different structures of relationship between alternative and regular medicine. See, for example, Matthew Ram-

sey's helpful study in "Alternative Medicine in Modern France," *Medical History* 43 (1999): 286–322. For an engaging survey of national differences in mainstream therapeutic practice, see Lynn Payer, *Medicine and Culture: Varieties of Treatment in the United States, England, West Germany, and France* (New York: Henry Holt, 1988).

49. My students are always impressed by a study reported some years ago that demonstrated that a widely utilized arthroscopic procedure for reducing the pain and immobility of osteoarthritis was no more effective than a sham in which surgeons pretended to operate on sedated patients. See Gina Kolata, "Arthritis Surgery in Ailing Knees is Cited as Sham," *New York Times*, July 11, 2002. It should be clear that I use the term *placebo* as a proxy for a larger cluster of emotional and psychosocial effects of the caregivers' diagnostic and therapeutic performance. See Anne Harrington, ed., *The Placebo Effect: An Interdisciplinary Exploration* (Cambridge: Harvard University Press, 1997), especially Howard Brody, "The Doctor as Therapeutic Agent: A Placebo Effect Research Agenda," 77–92.

50. Ted J. Kaptchuk, letter to the editor, *Annals of Internal Medicine* 132 (2000): 675. Contrast this with Ted J. Kaptchuk and David Eisenberg, "The Persuasive Appeal of Alternative Medicine," *Annals of Internal Medicine* 129 (1998): 1064, which also underlines the ways in which alternative and regular medicine necessarily overlap.

51. I have argued elsewhere for the efficacy of traditional Western therapeutics—that purging, bleeding and the like "worked" in a social and presumably physiological sense. See note 15 above and John Harley Warner, *The Therapeutic Perspective: Medical Practice, Knowledge, and Identity in America, 1820–1885* (Princeton, N.J.: Princeton University Press, 1997).

52. "My personal association with the culture of the West does not allow me to give to other forms the same value as to Western medicine. . . . " Owsei Temkin, *Double Face of Janus and Other Essays in the History of Medicine* (Baltimore: Johns Hopkins University Press, 1977), 16. I have been struck in recent years by attempts to study the "efficacy of prayer" with the validating tool of randomized clinical trials. Aside from the doctrinal issues of implying that an infinitely powerful and all-knowing deity would be bound by such contractual constraints, there remains the irreducible question of defining efficacy. Prayer, for example, provides an occasion to reflect on the meaning of disease and thus the place of humankind in a world of seeming randomness. What is the metric to weigh the benefit of a spiritual equity?

-- -- -- -- -- -- -- -- -- -- -- -- -- -- -- -- -- --

CHAPTER 8

HOLISM IN TWENTIETH-CENTURY MEDICINE

Always in Opposition

"HOLISM" IS AN ELUSIVE BUT INDISPENSABLE term for those seeking understanding of twentieth-century medicine. Like the fog hovering over a port, it is a persistent and sometimes even dangerous reality, yet hard to pin down. Holism challenges precise definition yet represents a historically real sensibility. One thinks of such parallel terms as republicanism, liberalism, romanticism, or—in the realm of medicine—vitalism and reductionism. Like all these resonant, culturally ubiquitous, yet soft-edged terms, holistic conceptions of the body and of the body in society constitute a significant historical phenomenon—because particular individuals in the past accepted inclusive and integrative assumptions about nature and society, invoked them, used them to justify ways of thinking and feeling about the world as it was and as it ought to be.

The careful reader will note that thus far I have not even attempted to define this elusive subject. And it should also be pointed out that the majority of physicians and social thinkers articulating holistic points of view never used the term itself, which is, in fact, a twentieth-century novelty; it is not to be found in the first edition of the *Oxford English Dictionary*.[1] Moreover, most were not systematic thinkers and were little interested in philosophical issues; their uses of holism were largely implicit and issue oriented. In any case, however, the task of definition is difficult because twentieth-century medical holism has to be understood primarily in terms of what it was not, and that is the mechanism-oriented reductionism that has come increasingly to characterize Western medicine during the past century.

Holistic thinking has drawn its moral edge, intellectual content, and occasions of social use from its oppositional place in a powerful and widely assumed narrative of medical progress. Holism has been a specter at this self-congratulatory feast, warning that progress has incurred costs as well as provided benefits. Holistic ideas have found their social meaning in antagonism to the bureaucracy, specialization, and fragmentation as well as the (related) reductionism that have come to dominate twentieth-century medicine. Medical holism, on the other hand, implies an emphasis on inclusiveness and integration, on pattern and interdependence—and a symmetrical disdain for the sufficiency of elucidating and manipulating discrete mechanisms in managing the healthy or diseased body.

The central components of holism have, of course, a much older history, and in this history inclusive and integrative notions have not always played an oppositional role. Holism is a fair description of the central and unquestioned worldview of medicine in the two millennia preceding the mid-nineteenth century, when medicine had organized itself around a conceptual world of aggregate outcomes and inclusive variables. But twentieth-century notions of holistic integration have been necessarily different from their classical and early modern predecessors. Meaning and nuance vary widely from place to place and from generation to generation; such explanatory schemes are necessarily context specific. Vitalism in the Montpellier of 1800 could not be the same as clinical holism in the Montpellier of 1930, even if one can point to seeming continuities of form and content.[2] Too much had changed in our understanding of the body.

By the mid-nineteenth century, at least a few self-confident advocates of a mechanism-oriented reductionism had already begun to question traditional notions of vital integration and the mystic specialness of living things.[3] In succeeding generations, such antivitalism grew even more pervasive. In the late-nineteenth and early-twentieth centuries, confidence in the laboratory and its ultimate promise grew inexorably in medical circles; reductionist values and assumptions became increasingly pervasive and unquestioned, almost self-evident. They seemed, in fact, no more than a necessary corollary of a steadily increasing knowledge base. An ever-more-intense focus on the cellular level and a growing biochemical and bio-physical understanding of the body promised eventual insight into the most intimate and fundamental mechanisms that constitute life. Traditional emphases on the organism's unity, integration, and interdependence could no longer be maintained without modification.[4] But they were maintained—by some physicians and biologists all of the time, and almost all some of the time.[5]

Integrative notions about the body, medicine, and society evolved in parallel with their increasingly powerful and hegemonic antagonist, a pervasive other against which integrative notions had necessarily to be defined and articulated. The germ theory, for example, structured and induced a skeptical opposition as it created an enthusiastic and often uncritical constituency. The formal content that expressed and legitimated such skepticism—the emphasis on predisposition and the environment—was not new, but the self-consciously oppositional and often defensive tone was.[6]

We are describing not only an abstract academic debate but also an array of related choices among real policies and particular career choices. In twentieth-century medicine, the invocation of holistic values has implied the polarizing antinomies of bedside as opposed to laboratory, of art as opposed to science, of the patient as individual and member of society in contradistinction to the patient as an aggregate of discrete mechanisms.

This oppositional role played by holistic ways of conceiving the world cannot be understood in terms of medicine alone. The development of medicine as a highly technical, elaborately institutionalized, impersonal, and bureaucratic system paralleled and in part constituted a more general shift in social forms and values. Medicine is not only a constitutive part of the twentieth century's highly institutionalized and impersonal social system but can also be used as a metaphorical tool for conceptualizing and criticizing that system—a particularly resonant microcosm of a narrowly technological, bureaucratic, and dehumanizing social macrocosm.

In the past century, notions of holism have reflected and communicated widespread anxieties about the structures and values of modern and postmodern society. We are all familiar with jeremiads against fragmentation and alienation, against the reductionism of market-oriented social relations—or the atomization and spiritual deadening nurtured by liberal individualism. Both left and right agree: things have fallen apart. Yet how does one go about restoring a more coherent and less alienating society? One recurring strategy has been the call for renewal of the social ties that—presumably—characterized a simpler organic community. Invocations of the need for social connectedness by critics and commentators have paralleled and reinforced medical emphases on the need to see the body as a unified system and to treat the patient as a whole person.[7]

FOUR HOLISMS

Holism is probably too general a term to be entirely useful; it includes a variety of styles and emphases that can and should be distinguished. I would suggest that there are four such conceptual styles, which I will refer to as historical, organismic, ecological, and worldview holism. Obviously there is something artificial about such abstracted categories. They have never been intellectually consistent and self-consciously maintained positions, but rather tendencies, emphases, lines of argument structured as much by moral and social agendas as by the logical demands of formal content. And they have been configured by particular historical actors in often inconsistent and overlapping forms. Each of these styles of argument is almost always associated in the thought of particular individuals with one or more of the other emphases. Nevertheless, it is important to delineate these distinctive styles of holism, to subject them to what might be seen, ironically, as a reductionist and abstracted style of analysis—that is, to disarticulate the parts in an effort to better understand the whole. A brief enumeration follows of some of the social contexts in which holistic ideas have been articulated. Finally, there is a discussion of a variety of contemporary trends, some pointing toward a softening of the conflict between holistic and reductionist styles of medical understanding, others implying quite the opposite.

HISTORICAL HOLISMS

There are two distinguishable forms of historical holism. The first may be referred to as metahistorical holism—in which health and disease are seen in the perspective of humankind's distant biological past. It is a holism of the human body in deep time, in which normative lessons about disease prevention and pathogenesis are drawn from speculative models of prehistoric biological and social development. Every aspect of regimen and behavior has in the past century been construed in terms of its consistency with the species' origins as hunter-gatherers and, subsequently, subsistence agriculturists. The acceptance of explicitly Darwinian ideas in the second half of the nineteenth century only underscored and revitalized the seemingly unproblematic nature of such assumptions—which had manifested themselves much earlier in primitivistic assumptions concerning the physical and moral advantages of a simple life. Human teeth, for example, as well as stomach and intestines were not designed for the soft, artificially spiced, and unnecessarily abundant diets to which civilized men and women had grown accustomed. Effete "overcivilized" women, unaccustomed to the strengthening routine of physical work, experienced pain and mortality in a natural process of childbirth that was undergone easily—even casually—by their sisters in earlier stages of civilization and in more primitive circumstances. Evolution provided, in other words, a newly compelling rationale for older cultural certainties. It situated the living body in a transcendent and value-confirming framework, a still living past from which lessons at once moral and physiological could be drawn. Not surprisingly, the late-nineteenth and twentieth centuries have seen a persistent interest in what might be called the evolutionary biology of disease and immunity (as well as persistent efforts to link and conflate the cultural and biological in speculative models of change over time). Most such efforts to place health and disease in the context of evolution were not meant as categorical rejections of mechanism—in some cases quite the contrary—but an effort to place those mechanisms in a larger integrative framework.[8]

A similar argumentive and rhetorical strategy could be used to comment on the linked physical and moral debility engendered in the course of Western society's development within the briefer compass of written history—a strategy that I tend to call "world-we-have-lost holism."[9] The presumed character and values of traditional village society have repeatedly served to constitute, that is, a compelling *ought*—a model of and for a morally and physiologically healthy polity. It is not surprising that "diseases of civilization"—of the sedentary, the scholarly, the anxious, and the affluent—should have been the concern of a variety of medical writers from the eighteenth to the late-nineteenth and twentieth centuries.[10]

A second kind of historicism—in this case specifically medical—was also used frequently in holistic arguments: the rhetorical deployment of

Western medicine's oldest and most honored intellectual tradition, a strategy often termed neo-Hippocraticism by nineteenth- and twentieth-century practitioners and historians alike. Such references conjure up a prestigious lineage of medical thought, one centered on the physiological uniqueness of the individual, and on an aggregate, eclectic, and inclusive understanding of health and disease.[11] This tradition could also underwrite another dimension of holism: the eclectic environmentalism summarized in the title of the Hippocratic treatise "Airs, Waters, and Places." Ultimately both aspects—the individual and the environmental and thus collective of the classical tradition—were consistent. Individual (and especially chronic) disease could be understood as the aggregated outcome of a variety of factors over time, whereas an instance of epidemic illness could be seen as the outcome of interactions between unique environmental circumstances and particular predisposed individuals. This explanatory tradition, which seemed firmly anchored in the texts and teachings of classical antiquity, provided in fact a usable and prestigious past that could be powerfully invoked by individuals uncomfortable with either the clinical or public-policy implications of reductionist medicine. For clinicians beset by the laboratory's and the specialist's diagnostic and therapeutic claims, it underlined the conceptual legitimacy of the multifaceted bond that integrated physician and patient. For those concerned with the public health, it pointed toward an active and eclectic concern for the environment.

ORGANISMIC HOLISM

If historical holism focused on conceptions of the body in time, organismic holism turned on the body understood as functioning unit. This style of holism illustrated another instance of not-always-self-conscious intellectual continuity as well: that of vitalism and arguments from design. Before the twentieth century, the body was seen as a unified interactive system utilizing but ultimately transcending the innumerable mechanisms that make up that system—a whole, in other words, that is greater than the sum of its constituent parts and processes. This way of thinking about living things had been assumed by centuries of physicians, philosophers, and pedagogues and used to demonstrate the Creator's presence in the body's design. Secular versions of this organizational scheme were still widespread in the nineteenth century.[12] Life, the argument followed, could not be reduced to its disaggregatable biochemical and biophysical mechanisms. There was an intrinsic and almost mystical "wisdom of the body" that permitted its integrated functioning.[13]

Although these issues had obvious theological and philosophical resonance, from the clinician's point of view they performed a more mundane function. Such holistic notions could, as has just been argued, help underwrite the social logic of the practitioner's role by affirming his or her ability to manage the unique configuration of problems posed by a patient's individuality.[14] The key to holism's clinical relevance thus rested on the patient's biological particularity. The inclusiveness and additive quality of the classical conception of the body and its functions justified the clinician's emphasis on the continuity of relationships over time, and that continuity implied tracking the interaction between a patient's body and every aspect of its environment. Only a stable physician-patient relationship allowed for the necessary accumulation of such understanding and observation.

It was only to have been expected that many late-nineteenth-century clinicians remained skeptical of the germ theory's sufficiency. Even if one conceded a necessary role to a particular microorganism in the etiology of a particular infectious disease, it was not necessarily a sufficient role. The actual encounter with a particular pathogen could easily be seen as only one among a variety of factors that determined whether disease was to be an outcome of that interaction.[15] It was only natural for most late-nineteenth-century clinicians to have continued to credit traditional notions of predisposition and resistance—that they should, for example, have so frequently invoked the metaphor of seed and soil to explain lingering problems of differential susceptibility.[16] Disease and its treatment are necessarily individual, and yet that focus on the idiosyncratic could serve as a justification for incorporating an eclectic—and holistic—universe of potentially relevant variables in explaining the particular incidence of sickness and health.[17] The idiosyncratic and the holistic were, that is, logically and contextually linked. Narratives of pathogenesis turning on lifetimes of cumulative events—on idiosyncrasy linked to biography—were a natural resource in opposing an increasingly dominant reductionist medicine.[18] The patient remained an individual; the physician continued to be in some measure a humanist, a source of wisdom and insight, not simply the purveyor of laboratory findings.

This central dichotomy in the negotiation and public presentation of clinical ideals is exemplified in a familiar rhetorical contrast—that between treating the whole patient and treating a disease or organ. These polarized visions of medical care communicate a stark and didactic message, one forcefully embodied in the formulaic juxtaposition of the image of the wise clinician with that of the technologically blinkered specialist (and by implication with the fragmented care unavoidable in a system dominated by narrowly focused experts). These sharply contrasting and emotionally charged ideal types have become clichés of social commentary, routinely and repetitively invoked since the end of the nineteenth century.

ECOLOGICAL HOLISM

If organismic holism focused on the body as organism moving through time, ecological holism has focused on the body in a particular social and physical setting. This was and is a point of view congenial to public-health advocates and consistent with the position that has come to be called "social medicine."[19] Health resulted from social, as well as individual, harmony, the central argument followed, from a proper—and thus moral—fit between a population and its economic and social, as well as physical, environment.[20] To advocates of social medicine, disease and health had to be seen primarily at the collective level.

Western medicine has been influenced by a rich tradition of such etiological thought, ranging from the eclectic environmentalism of the eighteenth century to the social medicine of the twentieth. Thinkers and controversialists as diverse as Rudolf Virchow and Friedrich Engels, Henry Sigerist and Thomas McKeown have all seen social and material circumstances as a cause of ill health. Some ills might grow out of the body's innate design; much constitutional disease, for example, and certainly old age and its concomitant ailments. But many other ills, the argument followed, were socially determined, created by circumstances that human society could modify or reverse. In the nineteenth and early twentieth centuries, not surprisingly, public-health reformers placed particular emphasis on the dangers associated with urban and industrial conditions— inadequate sewage and waste disposal, poor ventilation, contaminated water and food supplies, dismaying factory and housing conditions.[21]

This concern with social and material circumstances—what I have referred to as "ecological holism"—was not logically inconsistent with the organismic but was, nevertheless, ordinarily advocated selectively by different individuals guided by very different social agendas. Social medicine implies an emphasis on groups in context, not on particular individuals in particular clinical interactions. Similarly, there has been a conceptual and rhetorical affinity between individuals articulating a historical holism and those foregrounding an ecological point of view.[22] But again, as in the case of physicians and publicists concerned primarily with the individual body and its functioning (organismic holism) who also avowed a concern for the environment in which that body functioned, there is a sharp difference between foreground and background, between casual acceptance and programmatic centrality. Ecological holists shared a very different set of priorities from their organismic peers. For activists and advocates, logical consistency was less important than social commitment in the construction of their critical points of view.

WORLDVIEW HOLISM

The body plays a different role in this style of holism. It is a source of metaphor and value—a tool for thinking about society generally and its state of health or illness. When applied to larger social groups, body and disease metaphors are necessarily holistic, for they treat society as a unified whole—as an integrated and interconnected entity, a body writ large. And such thinking can be seen to work equally well in reverse. States of society can be seen as contributing to dis-ease in the bodies of individual men and women. Society itself can be pathogenic, the argument follows, the ills of its citizens indicators of a more general malaise. In both cases—in the rhetorical move from body to society and from society to body—the suggested relationship can articulate and legitimate cultural criticism.

Critics of contemporary medicine have been able to invoke a didactic parallelism between a reductionist biopathological understanding of the body and an anomic modernity, which produces and constitutes its own sort of spiritual sickness. Values and institutions have fallen apart, the argument follows: bodies into organs and nosological categories, medicine into procedures, market transactions, specialties, and hospitals. Holistic notions of value and function fit well into this metanarrative, a morality play directed at modern society and its costs. The master narrative of antimodernism has been and remains an important, in fact omnipresent, manifestation of worldview holism in medicine.

But holism can serve a variety of other roles as well in thinking about society and medicine. Worldview holism provides a value-laden framework—a higher knowingness—in which and from which the achievements of reductionist medicine can be made to appear insignificant and illusory. The particular frameworks can differ a great deal, but what each has in common is a claim to transcendence. From the spiritual viewpoint of an Ivan Illich, for example, the value of pain, the acceptance of death, the ability to order the material and temporal in a timeless framework remains central. In the secular moral philosophy of the laboratory scientist René Dubos, to cite a seemingly disparate yet in this sense parallel example, the anticipated celebration of medicine's ability to conquer disease is seen as a seductive "mirage"—illusory when placed in a larger context of biological evolution and adaptation. What Illich and Dubos have in common is, in fact, what I have called "worldview holism": they articulate intellectual and moral standpoints from which the material achievements of reductionist medicine can be understood as transitory and superficial. The self-confidence of reductionist medicine can be made to seem mere hubris, a short-sighted celebration of small equities acquired at great ultimate cost.[23]

Not surprisingly, worldview holism is routinely oppositional when confronting contemporary health-care systems in the West and the technical achievements that have done so much to legitimate and constitute them.[24] The seeming reality—the *is*—of an overly self-confident and unreflective technical medicine has often been contrasted with the *ought* of a medicine responsive to the emotional needs, the biological individuality, and social circumstances of particular men and women.

If the intrinsic needs, nature, and—by implication—rights of humankind can be used to justify a set of de facto goals for an ideal health-care system, then images of the body could legitimate a moral and organizational model for the larger society. Likening society to a body in need of balance, unity, and equilibrium has always provided a powerfully resonant framework for social commentary and criticism. One thinks immediately, for example, of the hypothetical social pathology of National Socialist ideology, which saw the Jews as pathogens, infecting an otherwise healthy social body, but this often-cited example is only an acute manifestation of a chronic ideological reflex; the use of health and disease metaphors in social discourse has been ubiquitous in Western society (as has the use of disease metaphors in dramatizing states of spiritual health; diseases of the soul as well as the body politic have been metaphorically invoked for centuries). One thinks as well of eugenic notions of race suicide and racial deterioration in a variety of societies that saw biopolicy as key to a vision of enhanced national power rooted in a morally committed citizenry.[25] Holistic notions of the body have had, as we are all well aware, a long and tenacious history as a rhetorical resource in the articulation of cultural and political discourse.

SITING MEDICAL HOLISM

Thus far I have sought to categorize the chief styles of holistic argument in twentieth-century medicine. Anyone familiar with the medical literature of the past century will concede that these arguments were used widely, repeatedly, in a variety of configurations, and for a variety of purposes. But used by whom? Which individuals and groups found these ideas plausible and useful and in which contexts were they most likely to be employed?

One clue lies in the social position of the individuals articulating them. This does not mean succumbing to the historian's special version of reductionism: mechanically associating particular ideas with particular social interests—and then connecting the dots. Ideas do not reduce themselves neatly to instrumental functions and social locations. Many factors interact to shape the likelihood of a particular individual finding holism congenial—and one of those factors is obviously the personal and idiosyncratic. Another is social location, and there are in fact, at least a half-dozen twentieth-century medical contexts in which such ideas were routinely deployed.[26]

One was the elite among consultants and teachers, from whose ranks clinical holism attracted a visible and voluble group of advocates. During the first third of the century, many elite practitioners were dismayed by what they saw as the proliferation of a narrow specialism and—related to that development—a slavish bedside subordination to laboratory findings. Diagnosis, the argument followed, had to be based on the context provided by the whole patient as seen over time.[27] No laboratory report could replace the unique and multidimensional understanding that grew out of a long-term physician-patient relationship. That mystical diad had to remain at the center of the physician's practice; medicine demanded an awareness of idiosyncrasy. Even though modern medicine was based on the data, techniques, and generalizing orientation of science, the clinicians' intuitive and inclusive approach could and should not be replaced by the laboratory's brutally one-dimensional and often misleading findings. This elite way of thinking about the physician's role did not imply a categorical rejection of science; quite the contrary. It did, however, prescribe the ideal clinician's attributes as incorporating the precision of science, the sensitivity of the gentleman, and the inclusive social understanding of the patient-oriented practitioner. The relationship between physician and patient should be just that—a relationship—and not an aggregate of isolated interactions.

Clinical intuition sharpened by years of experience had traditionally constituted a central aspect—and legitimation—for the elite consultant's social identity. Humanistic learning, like clinical wisdom, was also historically allied with assertions of status: an interest in history, the classics, and rare books was a conventional part of the gentlemanly practitioner's social self—the visible external stigmata of a reassuring inner wisdom.[28] No amount of earned status, no amount of scholarship and publication could entirely replace these ineffable qualities.[29]

In the course of the twentieth century, however, this gentlemanly orientation has become increasingly uncommon. Sensitivity to clinical idiosyncrasy has been gradually superseded by the accumulation of specialized technical skills—and the status associated with scientific understanding, the mastery of technical procedures, and the occupancy of hospital and medical school positions. Hospitals became the primary sites for medical care and their clinical laboratories and imaging departments the ultimate arbiters of diagnosis.

In response to these developments, holism has not died but adapted. Although the older class of genteel consultants is no longer dominant, many of their arguments have been refurbished and used by a variety of critical spirits in the world of medicine and the social sciences—some of them in fact specialists themselves concerned by an excessive subspecialization and the dominance of procedure-oriented medicine. New model holisms have reaffirmed in altered guise the traditional emphasis on an inclusive and integrative clinical sensibility. Critics of contemporary reductionism, impersonality, and intensely subspecialized practice have articulated a variety of such holistic positions in the past generation. George Engel's "biopsychosocial" approach provides an example well known to American clinicians,[30] but students of late-twentieth-century medicine are

familiar with a number of other reformist strategies. The distinction between illness and disease, for example; the emphasis on the patient as family member; even the emergence of medical school programs in humanities
and bioethics all recall—while reinventing—aspects of a traditionally inclusive holism.[31] Whatever their particular emphasis, that is, these arguments all constitute analytic and rhetorical strategies for questioning the
power and efficacy of a fragmented, reductionist, procedure-oriented
medicine.

Holistic thinking has been even more dominant in the world of
twentieth-century nursing. Since the era of Florence Nightingale, in fact,
holistic ideas of health, healing, and disease have been central to nursing
self-conceptions. Physicians cure, as the cliché goes, and nurses care. Op-
erationalizing the commitment to care has always involved an eclectic and
multidimensional understanding of the patient's environment—emotional
as well as physiological, in clinic and hospital as well as home. In some
ways the nursing profession's acceptance of holistic thinking is that profession's central ideological tradition. Reflections on nursing's defining
commitment to the whole patient, to care not cure, to the person as a
whole and not to a disease or organ are everywhere to be found in programmatic statements about nursing's social role and identity.

In the hands of nursing professionals, however, holism has a necessarily different meaning than when it is employed by physicians. It
communicates a location-specific meaning, one that reflects nursing's
structural—linked yet subordinate—relationship to medicine,[32] for expressions of nursing holism are, as we have seen, tied to professional status and clinical location. Nursing is by definition *not* medicine. Thus
nursing assertions of holistic caring are, unavoidably, a way of defining
the profession's relationship to medicine as much as a statement about
clinical aims and bedside practices. The meaning and specific content of
nursing holism are thus rather different from the holisms articulated by
clinician critics of an impersonal and specialized reductionist medicine.

The same can be said of the holism articulated by advocates of social
psychiatry and psychosomatic medicine, another source of twentieth-
century antireductionism. Advocates of mind-body holism have been a diverse lot, including psychologists and social scientists, as well as psychiatrists and public-health workers. But what this diverse group has in
common is an interest in system and process over time as well as links be-

tween the emotional and the somatic.[33] These psychiatric holisms are, of course, tied to an emphasis on body-mind relationships—and thus to interactions between the internal and external environments, between the family's microenvironment and society's macroenvironment—and between the physiological and psychological. The self-styled psychosomatic movement of the mid-twentieth century has played a prominent role in articulating a body-mind holism, incorporating emotional and cultural factors in a multifactorial model of disease causation.[34] Social psychiatry with its emphasis on groups and the social and political environment can be seen as allied historically and politically with ecological holism; both emphasize the pathogenic aspect of particular environmental situations.

To many laypeople, of course, medical holism appears most ubiquitously in the rationale for one or another lay-oriented, "antiestablishment" health system—almost all of which have developed out of and incorporated much older notions of symmetry between health and prudent regimen. From the nineteenth century to the present, alternative healers have asserted and reasserted many of the central components of medical holism. Certainly in water cure and its descendant naturopathy, in homeopathy, in a variety of Asian health systems—as in early modern Western medicine—emphasis on diet, environment, and regimen, for example, were framed and rationalized by speculative pathologies underlining the body's interactive dependence on lifestyle and environment.[35] The particular intellectual histories of such alternative healing systems cumulatively illustrate the cultural tenacity and explanatory power of holistic ideas as well as their widespread cultural diffusion.

And holistic ideas are, not surprisingly, also congenial to public-health and social-medicine advocates. One thinks for example, of Physicians for Social Responsibility, of advocates for urban public medicine, for education and drug rehabilitation as resources in the fight against AIDS. In the tradition of a much older social medicine, they have focused in an inclusive way on the social and political environment as potentially pathogenic. The proper role of medicine, the argument follows, is necessarily broader than that of intervening in the course of a particular illness or trauma, and that felt illness cannot be thought of as defined and constituted by a discrete biopathological mechanism. Nor can the responsibility of social medicine be limited to urging or requiring individuals—as individuals—to change their lifestyles. If society would prevent disease, it must, as a collective, address itself to cultural perceptions and to basic social and economic relationships—for in their interactive aggregate such fundamental realities help shape the organism's biological environment, as they do its social one.[36]

In the past quarter century, moreover, anxieties about the moral and emotional—as well as economic—costs of reductionist medicine have everywhere increased. One response has been the creation of a new discipline in the form of bioethics; another, the programmatic statements of would-be reformers of medical education who have, since mid-century, inveighed against a curriculum dominated by a narrowly technical subject matter. A variety of critics have, in other words, argued that medicine must transcend the technological ethos that assumes what can be done should be done—that sees entitlement to health care in technological terms alone and fails to account for the social, moral, and economic implications of that narrow but intensely focused vision.[37]

THE NEW HOLISM

The past century has witnessed a seemingly laudable and inexorable record of laboratory achievement in medicine. Occasional perturbations—the influenza pandemic of 1918, thalidomide, or AIDS, for example—have posed only momentary qualms. So, in a different way, did the worldwide depression of the 1930s, which underlined the disparity between medicine's technical capacity and the unequal provision of access to those healing resources. But to most laypeople and to their physicians as well, the history of medicine in the twentieth century was in essence a linear narrative of an inexorably increasing mastery over the body and—in clinical settings—an ability to understand and intervene in a variety of ills.

The mid-twentieth century seemed in fact to mark a discernible step forward in attaining that healing capacity. Antibiotics, cortisone, war surgery, and wound management all seemed to promise and constitute an ever-more-efficacious medicine—one based on the ability to understand and manipulate discrete biological mechanisms.[38] At the end of the century, though, a variety of physicians and social commentators voiced holistic second thoughts, questioning the expansionist optimism of a still-dominant reductionist—and acute care–oriented medicine. Such reservations are hardly universal and consistent but have manifested themselves in a revealingly eclectic variety of contexts.

The sources of this disillusion are obvious. Perhaps most fundamental are the unyielding realities of demography; the "laboratory's miracles" have yet to alter the limits of human biology; the conquest of infectious disease has only increased the burden of age and chronic illness. As demographically informed planners have warned since the 1920s, increasing life spans and the prevalence of chronic disease implies the shift of social resources toward care and management—toward a multidimensional consideration of the patient as person, including his or her social, economic, and emotional status. Clinical holism in the early twenty-first century is in good measure a recognition and articulation of such intrusive social realities. The episodic and mechanism-oriented style of medical practice seems increasingly one-dimensional, as unresponsive to questions of quality of life as it is recalcitrant to economic constraint. Like life itself, chronic disease poses continuing, unpredictable, and multidimensional demands. Accepting the need in so many instances for managing—not curing— illness, places the goals of medicine in a holistic, multidimensional framework. Pain and disability pose emotional, social, and economic, as well as physiological, problems. The illusion of neat, episodic, and technically defined solutions seems decreasingly viable.[39]

At the end of the twentieth century, moreover, a new pathological phenomenon underlined the continuing relevance of an eclectic and inclusive epidemiology. I refer to AIDS. Perhaps most obviously it underlines the way in which cultural values shape our response—society's imputation of meaning—to illness. Less obviously, perhaps, the still brief history of AIDS has made all of us aware of the way in which the modern laboratory's ability to explain a pathological mechanism has pointed ironically toward the need to see the intricate relationships among biological, ecological, and cultural factors in understanding the origins of the disease. Speculative etiologies of AIDS have repeatedly associated economic change, physical geography, and ecology with the linked natural history of a variety of animal species (relationships long of concern to parasitologists and epidemiologists if not to most clinicians and health planners). It has become, in fact, fashionable—if premature in detail—to see the emergence of AIDS and other new viruses in the perspective of a global ecology. Most important, the history of AIDS has underlined the arbitrariness of conventional distinctions between the cultural and biological; the disciplinary boundaries historically separating sociology, ecology, and biology dissolve as we seek to understand the etiology and epidemiology of AIDS and design preventive and therapeutic strategies.[40]

Related as well is the way we have come to understand health care—as symbolized by the way in which the three words "health," "care," and "system" have become in everyday usage almost a single word. And the concept of system implies inclusion and interaction, even if most commentators have focused narrowly in recent years on the economic aspects of that system. But critics of late-twentieth-century medicine have become aware that every aspect of society—of the state's responsibility for health care, for example, and of individual culpability for sickness—needs to be understood as relevant aspects of medicine; as much so, let us say, as models of normal physiological function and associated biochemical indicators. How we treat the sick and their families and what we agree upon as legitimate and blameless illness are as much part of our health-care system as medical school curricula and numbers of available hospital beds and facilities for laboratory diagnosis.[41] Holistically inclined critics have urged for generations that medical care be seen in terms of social commitment and ongoing relationships, and not simply as an aggregate of discrete, technically defined and legitimated market transactions. The world of medical care has, in other words, become a great deal more complicated since the end of the twentieth century. Every country in the Western world is, in some way or another, implicated in the same dilemma: providing care for an aging population in a world of ever-expanding technological options. It is impossible, as we are learning, to disaggregate the technical from the political, the economic, the cultural, and the demographic.

CONCLUSION: THE PARADOXES OF HOLISM

The more one looks at twentieth-century holism, the more elusive it becomes, the more it dissolves and reconfigures itself into its opposite. Even today in the early years of a new century, as so many individuals and groups question the efficacy of a previously unquestioned reductionist medicine, reductionism remains the dominant framework for thinking about health and disease. Without constitutive—and thus explanatory—mechanisms, for example, holism is at risk of being disdained as mysticism. It is hard to make the case that an organism is something more or different from the sum of its reductions; to do so is to guarantee dismissal by the great majority of the medical and scientific community. Mainstream holism ordinarily assumes the guise of an attack, not against a reductionist search for mechanism but against a premature and unsophisticated reductionism. Using reductionist means to justify and explain holistic ends constitutes, in other words, the holists' most persuasive rhetorical strategy in the early twenty-first century. But it is not a very effective one.[42]

A particular phenomenon can, moreover, often be construed in *either* reductionist or holistic terms, depending on one's level of analysis and point of observation. Foreground can become background, background foreground. There are many such examples in the history of twentieth- and twenty-first-century medicine. How, to cite a striking example, does one regard pellagra? Does the discovery of the dietary origins of this disease constitute an impressive example of an elusive and protean clinical problem being explained—and banished—through the elucidation of a discrete biopathological mechanism? This is certainly one way of construing pellagra's twentieth-century history. Yet one can also see the incidence of pellagra as the peculiar outcome of historical and social circumstances— in which the dietary cause once understood serves (like our pinpointing the infectious agent responsible for AIDS) as a sampling device illuminating the complex of events and relationships that allowed pellagra to become a social reality.

AIDS provides, in fact, a dramatic parallel case. From one point of view, it presents a de facto brief for the laboratory's mechanism-oriented way of knowing. From another, as we have seen, it presents an argument for an inclusive and integrative way of thinking about disease and the society in which it occurs. Understanding the virus's mode of reproduction does not, in fact, explain the epidemic, or automatically confer insight into the ailment's idiosyncratic clinical course, much less an effective preventive strategy. Mechanism becomes then an aid to epidemiological understanding, allowing us to reconstitute the particular configuration of factors that allowed AIDS to become a visible medical problem. From one point of view, in other words, the mechanism is the reality; from another it is an indicator, a tool for unpacking complex social and ecological relationships.

In terms of their ideological uses, moreover, medical and biological holisms are equally elusive. Holistic arguments have no necessary and predictable role in social discourse; such ideas can be deployed by the left as well as right, the defender of a status quo as well as its would-be antagonist. Holism can have an individualist, atomistic, and antinomian streak—glorifying the individual's multidimensional needs and perceptions as the basic building block of society—or it can, as we have suggested, justify an authoritarian, centralized view of the state in which the autonomy of the individual is minimized by a kind of mystic antirationalism. In the understanding of medicine as clinical practice, holism can be profoundly traditional in its ability to legitimate the individual and idiosyncratic physician-patient relationship as at once the moral and intellectual center of medicine, but it can also be used to justify a centralized and bureaucratically rationalized medicine.

The conflict between idiosyncrasy as opposed to the reductionist—schematic and generalizing—tendency is central to medicine's history, from the formation of nineteenth-century ideas of disease specificity to its modern American incarnation in the diagnosis-related group, the nosological table, and psychiatry's diagnostic and statistical manual. Although there is an essential tension between the two ways of understanding medicine, they are, in practice, mutually constitutive. Diseases must always manifest themselves in the bodies and minds of individual patients. And—reciprocally—a focus on the individual and the idiosyncratic leads the physician toward the general and the schematic as he or she seeks to treat and predict. The impersonal and reductionist understanding of disease—the naming of an illness, the imposition of a future trajectory, often the specification of a standardized treatment—demands a linkage between the individual and the collective, a framework for assimilating the incoherence and arbitrariness of individual experience to a collective meaning. At the individual level it provides a framework of agreed-upon meaning for physician and patient.

Holism and its reductionist antagonists may remain in opposition, but they will not disappear. They cannot because they are part of an interactive fabric of understanding, one integrating the individual and society. The conceptual constraints we often resent—the abstract and categorical understanding of disease, pathological mechanism, and standardized treatment—represent at the same time a reassuring source of structure and coherence in a world otherwise beset by randomness.

Medical thinking has been historically organized around a number of mutually constitutive oppositions. Just as the opposition between art and science represents an ironically tenacious marriage, so does the conflicted relationship between the inclusive, organic, and holistic way of framing problems and the reductionist and mechanism oriented. There are no simple answers, no winner or loser in this debate, no dissolving this fundamental instance of unity in diversity.

NOTES

1. The letter "H" in the OED was prepared between 1897 and 1899. *Webster's New International Dictionary . . . Second Edition Unabridged* (Springfield, Mass.: G. & C. Merriam, 1936) defines holism as "the philosophic doctrine of General Smuts that the determining factors in nature, and particularly in evolution, are wholes such as organisms and not their constituent parts." Although as indicated in this definition, the word itself has been associated with the South African biologist and political leader J. C. Smuts, the conceptual burden of his neologism is, of course, much older. See Jan C. Smuts, *Holism and Evolution* (New York: Viking Press, 1961 [1926]).

2. Montpellier is well known to historians of the late eighteenth and early nineteenth centuries as a center of vitalistic physiology and pathology. See, for example, Elizabeth A. Williams, *The Physical and the Moral: Anthropology, Physiology, and Philosophical Medicine in France,* 1750–1850 (Cambridge: Cambridge University Press, 1994), 20–66. French clinical holism in the first half of the twentieth century focused on susceptibility and idiosyncrasy.

3. Perhaps most conspicuously—at least to historians—in the programmatic form articulated by a number of mid-nineteenth-century German physiologists and physicists.

4. Or at least without being articulated—paradoxically—in terms of the concrete integrating mechanisms through which that mystical overarching unity was constituted. I shall return to this irony in the concluding section of this chapter.

5. Most medical men and biologists were, that is, neither thoroughgoing reductionists nor self-conscious critics. Most simply assumed elements of both, but it is safe to say that in the course of the twentieth century reductionist approaches became inexorably more pervasive, especially at the level of elite practice and in the growing prominence of what have come to be called the basic medical sciences. Programmatic attacks on holism and vitalism seemed no longer necessary.

6. Relevant as well to an ever-more-pervasive laboratory-oriented reductionism—even in clinical medicine—was a medical acceptance of disease specificity that

antedated and was powerfully affirmed by the notion that specific microorganisms could be understood as causal in the incidence of particular diseases.

7. In twentieth-century medicine, holism of the left tended to present itself in antinomian, anti-authoritarian form—because reductionism is built into credentialism, into bureaucracy, into central as opposed to local sites of authority and decision making. Holism of the right tended to emphasize a spiritualized community identity and, at times, legitimated a mystical unity manifested in centralized value-embodying authority. One thinks, most dramatically, of the ideology and policies associated with German National Socialism.

8. The literature in this area is enormous, but representative of these speculative discussions are Sir John Bland-Sutton, *Evolution and Disease* (New York: Scribner & Welford, 1890); James T. C. Nash, *Evolution and Disease* (New York: William Wood, 1915); John George Adami, *Medical Contributions to the Study of Evolution* (London: Duckworth and Co., 1918); and Jan Danysz, *The Evolution of Disease with a Discussion of the Immune Reactions Occurring in Infectious and Non-Infectious Diseases: A Theory of Immunity, of Anaphylaxis and of Antianaphylaxis,* trans. Francis M. Rackemann (Philadelphia: Lea & Febiger, 1921). For a contemporary synthesis, see Paul W. Ewald, *Evolution of Infectious Disease* (Oxford: Oxford University Press, 1994). See also "Pathologies of Progress," chapter 5 in this volume.

9. I refer to the title (and content) of Peter Laslett's well-known *The World We Have Lost* (New York: Charles Scribner's, 1965).

10. There is an abundance of such latently homiletic writings, ranging from late-nineteenth-century concerns about diseases of civilization to contemporary warnings against "unnatural diets" as a cause of cancer. It might be noted that the possible objective validity of such concerns about dietary risk in oncogenesis does not invalidate the cultural function of such anxieties. See, among many scores of such texts, Benjamin Ward Richardson, *Diseases of Modern Life* (New York: D. Appleton & Co., 1876). On the perils of a sedentary life, see, for example, Samuel A. D. Tissot, *An Essay on Diseases Incident to Literary and Sedentary Persons,* 2nd ed. (London: J. Nourse and E. and C. Dilly, 1769); Daniel Newell, *The Guide to Health: Designed to Promote the Health, Happiness, and Longevity of Students and All Others in Sedentary Life, and Especially Invalids* (Boston: S. T. Farren, 1825); [Chandler Robbins], *Remarks on the Disorders of Literary Men; Or, an Inquiry into the Means of Preventing the Evils Usually Incident to Sedentary and Studious Habits* (Boston: Cummings, Hilliard & Co., 1825); and William Cornell, *A Series of Articles on Clerical Health* (Boston: Brown, Taggard & Chase, 1858).

11. There is a substantial literature turning on the distinction between this holistic (individual and physiological) orientation and reductionist notions that have in some measure displaced—and interacted with—it. See, for example, Owsei Temkin, "The Scientific Approach to Disease: Specific Entity and Individual Sickness," in *Scientific Change: Historical Studies in the Intellectual, Social and Technical Conditions for Scientific Discovery and Technical Innovation from Antiquity to the Present,* ed. A. C. Crombie (New York: Basic Books, 1963), 629–47 and N. D. Jewson,

"The Disappearance of the Sick-Man from Medical Cosmology, 1770–1870," *Sociology* 10 (1976): 224–44. I have tried elsewhere to deal with this persistent and dichotomous aspect of medical thought. See Charles E. Rosenberg, "The Therapeutic Revolution: Medicine, Meaning, and Social Change in Nineteenth-Century America," *Perspectives in Biology and Medicine* 20 (1977): 485–506, and "Explaining Epidemics," in Rosenberg, *Explaining Epidemics and other Studies in the History of Medicine* (Cambridge: Cambridge University Press, 1992), 293–304. All these citations can, of course, be construed themselves as part of an ongoing discourse surrounding the relationships among value, social organization, and epistemological conviction in medicine.

12. Evolution, it should be noted, was effortlessly made to provide a materialist recasting of natural theological arguments from design. I have discussed the widespread diffusion of such views in "Catechisms of Health: The Body in the Prebellum Classroom," *Bulletin of the History of Medicine* 69 (1995): 184–92.

13. The reference is to Walter B. Cannon's *The Wisdom of the Body* (New York: W. W. Norton, 1932). Although Cannon was much concerned to specify the physiological and biochemical mechanisms that constituted the body's wisdom, the very title of his book invoked older notions of design, and in its metaphoric anthropomorphism imparted a necessarily holistic volition and unity to the body as it functioned through time. Cannon also placed his schematized body in evolutionary time in a dramatic and self-conscious way. Physiological mechanisms could be construed in terms of their survival characteristics. See, for example, his influential *Bodily Changes in Pain, Hunger, Fear and Rage: An Account of Recent Researches into the Function of Emotional Excitement* (New York: D. Appleton & Co., 1915). For Cannon's life, see Saul Benison, A. Clifford Barger, and Elin C. Wolfe, *Walter B. Cannon: The Life and Times of a Young Scientist* (Cambridge, Mass.: Belknap Press, 1987), and Allen Young, "Walter Cannon and the Psychophysiology of Fear," in *Greater than the Parts: Holism in Biomedicine, 1920–1950,* ed. Christopher Lawrence and George Weisz (New York: Oxford University Press, 1998), 234–56.

14. In addition to twentieth-century holism in physiology and immunology, holistic conceptions were influential—as is well known—in interwar psychology and philosophy; again, the emphasis was not to dismiss mechanism, but to emphasize pattern, configuration, and integration. Psychological function was, that is, more than a sum of discrete mechanisms. See, in particular, Mitchell G. Ash's invaluable *Gestalt Psychology in German Culture, 1890–1967: Holism and the Quest for Objectivity* (Cambridge: Cambridge University Press, 1995); Anne Harrington, *Reenchanted Science: Holism in German Culture from Wilhelm II to Hitler* (Princeton, N.J.: Princeton University Press, 1996).

15. Predisposition and constitution were fundamental conceptual building blocks in traditional explanations of the occurrence of individual sickness, especially in chronic ailments. Clinical intuition questioned neither, and both could be integrated in a cumulative—if speculative—narrative. Immunology can, in this clinical and historical context, be seen in part as the inheritor of a persistent focus on

predisposition and constitution—acting in a kind of dialectic with the implicitly reductionist thrust of late-nineteenth-century bacteriology and its emphasis on the specificity of infectious agents.

16. Similarly, twentieth-century French commentators on such issues refer frequently to the problem of *terrain,* invoking the same metaphorical argument. See George Weisz, "A Moment of Synthesis: Medical Holism in France between the Wars," in Lawrence and Weisz, *Greater Than the Parts,* 68–93.

17. Consistently enough, adaptation became a key concept in late-nineteenth- and early-twentieth-century medicine, used to explain the occurrence of health and disease and provide an integrative and plausibly post-Darwinian understanding of pathogenesis in an increasingly reductionist intellectual environment. Successful adaptation to a particular circumstance—emotional or material—meant health; failure to adapt brought dysfunction and ultimate disease.

18. Holistic assumptions obviously fit more comfortably in explaining chronic diseases than in acute infectious ills—even when a specific microorganism could be associated with the clinical entity. Tuberculosis, for example, was a focus for the holistically oriented more frequently than, say, influenza; idiosyncrasy and environment both seemed to play a role in this omnipresent ailment's incidence and unpredictable clinical course. Ecological and social variables could also play a role— so that "social" explanations of disease incidence could have a collective as well as an individual aspect.

19. There is an elaborate literature on the history of social medicine. For the still classical introduction to this tradition, see George Rosen, "What Is Social Medicine?" *Bulletin of the History of Medicine* 21 (1947): 594–627. It is reprinted with a number of related essays in Rosen, *From Medical Police to Social Medicine: Essays on the History of Health Care* (New York: Science History Publications, 1974).

20. There is a coherent, if not always continuous historical tradition of what might be called ecological epidemiology. See, among many such examples, Erwin H. Ackerknecht, *Malaria in the Upper Mississippi Valley, 1760–1900,* Supplement to the *Bulletin of the History of Medicine,* no. 4 (Baltimore: Johns Hopkins Press, 1945); L. Fabian Hirst, *The Conquest of Plague: A Study of the Evolution of Epidemiology* (Oxford: Clarendon Press, 1953); John Ford, *The Role of Trypanosomiases in African Ecology: A Study of the Tse Tse Fly Problem* (Oxford: Clarendon Press, 1971); Alfred W. Crosby, Jr., *The Columbian Exchange: Biological and Cultural Consequences of 1492* (Westport, Conn.: Greenwood, 1972); John Farley, *Bilharzia: A History of Imperial Tropical Medicine* (Cambridge: Cambridge University Press, 1991); Ken De Bevoise, *Agents of Apocalypse: Epidemic Disease in the Colonial Philippines* (Princeton, N.J.: Princeton University Press, 1995).

21. Which is not to say that they ignored individual behavior and cultural practice—or even an older emphasis on climate and geography—in the construction of their inclusive style of epidemiology.

22. Note, for example, how Thomas McKeown, the influential advocate for preventive and social medicine, emphasized the "unnaturalness" of late-twentieth-

century diets and lifestyles in legitimating his position. See *The Role of Medicine: Dream, Mirage or Nemesis?* (Princeton, N.J.: Princeton University Press, 1979), 79–81. His emphasis on the divergence of humankind from the optimum lifestyle inbuilt in their genes can also be seen as a characteristic example of the post-Darwinian version of argument from design—as discussed above.

23. Contrast this with Ivan Illich, *Medical Nemesis: The Expropriation of Health* (New York: Pantheon, 1976); René Dubos, *Mirage of Health: Utopias, Progress, and Biological Change* (New York: Harper, 1959); Dubos, *Man Adapting* (New Haven, Conn.: Yale University Press, 1965). My reference to Illich implies the existence of a different realm of holism, the explicitly religious and mystical. In this chapter I have avoided discussion of this way of thinking about the world. Spiritual commitment is not explicitly a part of medical thought—even though it has been a fundamental component in the shaping of modem health-care institutions and a significant factor in the determining of individual medical careers and worldviews.

24. It should be noted that the very term *health-care system* that has grown so fashionable in the past generation draws on metaphors of function and interconnectedness.

25. French anxieties about a falling birth rate in the late nineteenth and early twentieth centuries are a particularly conspicuous example. English fears of national deterioration following the Boer War provide another well-known example.

26. And the medical groups that articulated these ideas were far from monolithic in their views. Some members of a particular group might use them enthusiastically; others showed only sporadic interest, whereas still others might explicitly disown them. Specialties, too, vary in their affinity for holistic ideas. And such ideas drawn from biology and medicine were, of course, used by a variety of lay publicists and social commentators—but I have limited my discussion here to the world of health and medicine.

27. See Christopher Lawrence, "Incommunicable Knowledge: Science, Technology and the Clinical Art in Britain, 1859–1914," *Journal of Contemporary History* 20 (1985): 503–20; Lawrence, "Moderns and Ancients: The 'New Cardiology' in Britain, 1880–1930," *Medical History,* suppl. no. 5 (1985): 1–33. See also Stanley J. Reiser, *Medicine and the Reign of Technology* (Cambridge: Cambridge University Press, 1978), for a synthetic restatement of an older discomfort with a technology-oriented clinical practice.

28. William Osler was probably the most prominent of such elite consultants in the Anglo-American world. His well-known interest in history, belles-lettres, and bibliography constituted another obeisance to the gentlemanly style in clinical practice. But it should be recalled that Osler's original fame grew out of work in clinical diagnosis and pathological anatomy, symbolizing the increasingly reductionist grasp of specific disease entities on the profession. A parallel argument can be made about Harvard and Boston's Richard Cabot, an eloquent and effective spokesman for social medicine but equally famous as a teacher of differential diagnosis. On Osler, see Michael Bliss, *William Osler: A Life in Medicine* (Oxford: Oxford Univer-

sity Press, 1999). Harvey Cushing's *The Life of Sir William Osler*, in two volumes (Oxford: Clarendon Press, 1925), remains an important, if hagiographic, source. By now it has become a historical document itself, an exemplification of an elite model of the ideal academic clinician.

29. Which is not to say that the publication of books and articles was not a significant aspect of the nineteenth-century consultant's status and self-image. Academic accomplishment was seen as a key aspect of the elite consultant's role, but not entirely sufficient in itself to constitute that role.

30. Engel's eclectic program would have seemed no more than the commonsense truth of clinical practice to any American or European practitioner before the mid-nineteenth century. What is particularly striking is its seeming novelty in the late-twentieth-century medical world. See George Engel, *Psychological Development in Health and Disease* (Philadelphia: Saunders, 1962).

31. Arthur Kleinman has been perhaps the most influential in elaborating the distinction between the physician-centered concept of disease and the patient-centered, experiential notion of illness. See, for example, Kleinman, *The Illness Narratives: Suffering, Healing, and the Human Condition* (New York: Basic Books, 1988). The necessarily antireductionist emphasis on narrative and experience has become widespread. See, for example, Kathy Charmaz, *Good Days, Bad Days: The Self in Chronic Illness and Time* (New Brunswick, N.J.: Rutgers University Press, 1991); David B. Morris, *The Culture of Pain* (Berkeley: University of California Press, 1991); and Kathryn M. Hunter, *Doctors' Stories: The Narrative Structure of Medical Knowledge* (Princeton, N.J.: Princeton University Press, 1991).

32. I should perhaps have said medical men—because the social meanings of medicine and nursing have historically reflected gender as well as status and authority. In a more extended context, this theme would have to be discussed at greater length because of the gendered resonance shaping our customary usages of care and cure, intervention and management, dominance and submission to the body's healing potential.

33. Twentieth-century dynamic psychiatry in its numerous manifestations is holistic insofar as it is oriented toward system and interactive process over time—even if the system it studies is often limited to the family; however, commentators have often underlined a comparative lack of interest in somatic interactions and the larger social and economic context. Critics of the dynamic style in twentieth-century psychiatry picture it—whatever its particular and varied manifestations—as a form of abstracted, theory- (if not laboratory-) dependent reductionism.

34. The emphasis on such connections is, of course, a fundamental and unquestioned aspect of traditional medical theory. I have discussed this at greater length in "Body and Mind in Nineteenth-Century Medicine: Some Clinical Origins of the Neurosis Concept," *Bulletin of the History of Medicine* 63 (1989): 185–97. There is an enormous literature in this area.

35. Contemporary fears of chemical pollution, for example, might be said to reflect and incorporate an aspect of these traditional notions, namely, the body's

dependence on an ongoing interaction with its environment. Disease could in the framework of this speculative pathophysiological model result from the aggregate intake of even small amounts of unnatural substances over time.

36. Parallel holisms have been invoked as well, we have already suggested, by some basic scientists, appalled at the narrowness of a mechanism-oriented, acute care–oriented medical profession. I refer to such scientists as René Dubos, for example, who, as we have seen, emphasized the adaptive nature of organisms and the ephemeral quality of many of medicine's twentieth-century "triumphs." Dubos's arguments are not political in the conventional sense but do have relevance to intra-academic status and values—and to the past generation's debates over health and environmental policy. For a valuable introduction to Dubos's thought, see Barbara Rosenkrantz, "Introductory Essay: Dubos and Tuberculosis, Master Teachers," in *The White Plague: Tuberculosis, Man, and Society,* ed. René and Jean Dubos (New Brunswick, N.J.: Rutgers University Press, 1987), xiii–xxviii. *The White Plague* was originally published in 1952.

37. I have sought to trace the development of this reductionist focus on mechanism-oriented technology in its relationship to the American hospital in "Inward Vision and Outward Glance: The Shaping of the American Hospital, 1880–1914," *Bulletin of the History of Medicine* 53 (1977): 346–91, and *The Care of Strangers: The Rise of America's Hospital System* (New York: Basic Books, 1987). See also in this connection Joel D. Howell, *Technology in the Hospital: Transforming Patient Care in the Early Twentieth Century* (Baltimore: Johns Hopkins University Press, 1995).

38. The cold war narrowed the parameters of debate. A scientifically based social rationality came to seem not only a part of the American way of life but a necessary attribute of the informed citizenship presumed to underpin that way of life. Even many physicians on the left who questioned the monolithic polarities of the cold war did not question the value and legitimacy of contemporary scientific medicine; in fact, they deplored the injustice of a society that could not make its newfound medical capabilities available to the less privileged. The problems of health care grew, they felt, out of the inequities structured into capitalism and were not inherent in the nature of mid-twentieth-century medicine.

39. Mirroring demographic realities, there has been a recent upswing in historical interest in the patients' experience in chronic disease. See, for illuminating examples, Barbara Bates, *Bargaining for Life: A Social History of Tuberculosis, 1876–1938* (Philadelphia: University of Pennsylvania Press, 1992); Linda Bryder, *Below the Magic Mountain: A Social History of Tuberculosis in Twentieth-Century Britain* (Oxford: Oxford University Press, 1988); Sheila M. Rothman, *Living in the Shadow of Death: Tuberculosis and the Social Experience of Illness in American History* (New York: Basic Books, 1994); Chris Feudtner, "The Want of Control: Ideas, Innovations, and Ideals in the Modern Management of Diabetes Mellitus," *Bulletin of the History of Medicine* 69 (1995): 66–90; Feudtner, "A Disease in Motion: Diabetes History and the New Paradigm of Transmuted Disease," *Perspectives in Biology and Medicine* 39 (1996): 158–70; Steven J. Peitzman, "From Bright's Disease to End-Stage Renal Disease," in

Framing Disease: Studies in Cultural History, ed. Charles E. Rosenberg and Janet Golden (New Brunswick, N.J.: Rutgers University Press, 1992), 3–19; Roy Porter, "Gout: Framing and Fantasizing Disease," *Bulletin of the History of Medicine* 68 (1994): 1–28.

40. The emergence of drug-resistant pathogens and the much publicized reemergence of ills such as tuberculosis have only underscored the relevance of larger biological and evolutionary perspectives—what I have previously described as historical holism. Insofar as we focus on the twentieth century's use and misuse of antibiotics, for example, we also make it impossible to ignore the arbitrariness of boundaries between the cultural, the historical, and the biological.

41. The exhilarating power of reductionist visions should not be underestimated. The lure of categorical and definitive mechanism-oriented solutions to clinical problems remains powerful, as demonstrated in contemporary hopes for stem cell research and biotechnology generally. See "The New Enchantment: Genetics, Medicine, and Society," chapter 6 in this volume.

42. Medicine has seen a variety of advocates of cultural and multidimensional pathologies arguing for the legitimacy of their hypothetical disease constructs in terms of particular mechanisms. How else could they convince their contemporaries to accept such novel points of view? See, for example, the works by Walter B. Cannon cited in note 13 above; Sarah W. Tracy, "George Draper and American Constitutional Medicine, 1916–1946: Re-inventing the Sick Man," *Bulletin of the History of Medicine* 66 (1992): 53–89.

CHAPTER 9

MECHANISM AND MORALITY

On Bioethics in Context

ONE CAN HARDLY IGNORE THE WIDELY SHARED conviction that we are living through a period of crisis in health care. And that crisis is more than economic and administrative—though its most egregious symptoms present in these interrelated forms. One need only pick up a newspaper or magazine to be reminded of the omnipresent and multidimensional nature of the problems confronting Western medicine. Many of these perceived dilemmas turn on rapid technical change and the difficulty of creating an institutional, economic—and moral—context in which these new clinical, policy, and research options can be managed. Not surprisingly, bioethics is often invoked—as both symptom and possible remedy—in discussions of these disheartening realities. How are we to think about this enterprise? Site it in social space? Understand its several interrelated identities? It is no easy task. Contemporary bioethics constitutes a particularly elusive and seemingly novel challenge for the historian. Value assumptions have always shaped medicine as a social enterprise, yet these values have often been implicit and unspoken, the moral common sense of each generation interacting with technical, professional, institutional, and economic factors to configure a place- and time-specific set of clinical realities.

I have always regarded medicine as being in one of its dimensions timeless—and sacred. Women and men become ill, feel pain, and seek care. Although it functions in the market, medicine cannot be reduced to a series of market transactions. It makes me uncomfortable to hear health care referred to as a product—and to see it advertised. I am dismayed by the ubiquitous use of the term *health-care system* when the presumably inclusive meaning of the term *system* is limited to bureaucratic and economic relationships. And I share much older—and recurrent—anxieties about the possible abuse of patients as teaching and research material. I feel, moreover, that my sentiments are neither idiosyncratic nor inappropriately nostalgic, but are an expression of a widespread, if not always clearly articulated, social consensus. The creation of that enterprise called bioethics is, I believe, in part a response to such feelings of disquietude. In a world of change-inducing technical capacity, of credentialed experts and bureaucratic structures, we have created in bioethics what might be called a "social technology," a novel entity configuring intellectual, attitudinal, and institutional elements and whose function it is to help rationalize and thus manage these disturbing realities. But like all technologies, its function and meaning are determined by their context of use. Bioethics transcends peculiarities of place and policy, yet inevitably reflects and incorporates those peculiarities. It is an ideal object of study for the historian or social scientist.

MEDICAL CARE AND SOCIAL OBLIGATION: THE AMERICAN CASE

Some markers of change in American medicine strike me with particular force. The *New York Times,* for example, reported that Montefiore Hospital had announced its intention of entering into a joint venture with a for-profit corporation; it planned to open a chain of twenty-four-hour cancer and HIV clinics. "The number one problem for not-for-profit institutions," the president of Montefiore explained, "is capital formation."[1] In Philadelphia, the Pennsylvania Hospital, America's oldest private general hospital, first sold its historically important psychiatric division to a for-profit provider, then sold itself—after an independent existence of a quarter of a millennium—to a rather more youthful entity called the University of Pennsylvania Health System—which announced its plans to send four "experts in 'clinical reengineering' to look for ways to make cost-effective changes in clinical care" at its new acquisition.[2] The Hospital of the University of Pennsylvania had just finished its own "reengineering."[3]

Particularly revealing among my collection of media indicators was an ironic—and enlightening—juxtaposition of stories on the front page of the *New York Times*.[4] In the upper right-hand corner was a report that National Institutes of Health funding was likely to be increased in next year's budget. Cancer could be understood and treated, it was explained. "We are in a golden age of discovery," the director of the National Cancer Institute contended, "one unique in human history. . . . Knowledge about the fundamental nature of cancer is exploding." Basic science was closing in on mankind's ancient enemy, and relentless Washington lobbying could be relied on to nurture this laudable enterprise. A coalition of interested parties—patient advocacy groups, physicians, and medical schools—had joined in supporting an effort to double the NIH budget over the next five years. "We plan a grass roots campaign inside and outside the Beltway," the president of their lobbying firm explained candidly: "It will be run the same way Northrop Grumman lobbies for the B-2 bomber." Immediately to the left of this relentlessly euphoric report of promised laboratory achievement was a background story on the emotional and physical pain associated with the multiple births resulting from contemporary fertility treatments: "joy and sorrow follow medical miracle," read one of the subtitles in this sobering overview.[5] Whether the placement of these stories on the front page of the *Times* was a compositor's whim or an implicit editorial comment, the message seems undeniable. Technology, market incentives, and public policy had changed and are changing every aspect of medical care, while society has been less than successful in anticipating the consequences of such change.

But expectations remain boundless. When commentators reflected some years ago, for example, on medicine in the new millennium, they dwelt overwhelmingly on technology. A special issue of *Life* (fall 1998), to cite one instance, was devoted to "Medical Miracles for the Next Millennium." The cover promised "21 breakthroughs that could change your life in the 21st century: Gene therapy/edible vaccines/memory drugs/grow-your-own-organs." Little attention was paid in the magazine's worshipful depiction of laboratory progress to the ironic and seemingly paradoxical growth of a widespread fear of that same technology's human implications. Similarly illuminating was an issue of *Time* on the "Future of Medicine." The subtitle promised to explain "how genetic engineering will change us in the next century." The striking cover illustration was a stylized caduceus, a snake's head morphing into a coil of DNA.[6] How better to symbolize medicine's changing and conflicted shape in a world of relentless laboratory progress and media-heightened public expectations? The cover's powerful visual metaphor represents as well two seemingly inconsistent yet mutually constitutive aspects of contemporary medicine: the technical and the sacred—the cultural power of laboratory novelty and the persistence of a self-conscious ethical tradition.

I would argue that this brief sampling of media reports provides a useful microcosm of a structural and emotional macrocosm. It illustrates not only a perceived crisis in public policy but also a fundamental inconsistency between values and expectations and the concrete social and economic relationships in which such convictions and perceptions are necessarily embedded.

The American health care system is marked, that is, by a characteristic disconnect: on the one hand, uncritical faith in the power of the laboratory and the market, on the other a failure to anticipate and respond to the human implications of technical and institutional innovation. And one of those dilemmas grows directly out of our expansive faith in technical solutions to clinical problems; as we are well aware, sickness, pain, disability, and death are not always amenable to clinical intervention. In the late twentieth century, such conflicts are both public policy issues and—inevitably—elements in individual physician-patient relationships. The question, of course, is relating the particular to the general, understanding the choices that face individuals in recurring social interactions— in some sense weighing and understanding degrees of individual autonomy, of professional and of collective social obligation. I would contend that bioethics must ultimately address such questions that are necessarily historical and unavoidably moral: the move from the individual to the social, from meaning to structure—and in terms of medicine, from the clinical encounter to the larger society in which that encounter takes place.

To a historian, many of the dilemmas that beset contemporary medicine—and which I have sought to illustrate—are strikingly different from parallel realities in previous American generations. The world of social value and thus obligation was very different, for example, when New York's Montefiore Home for Chronic Invalids opened its doors in 1884 and certainly when the Pennsylvania Hospital was established in the 1750s. One was Jewish in origin and management, the other Quaker, but they shared fundamental characteristics. Pious and paternalistic activism, the exchange of care for deference, was as central to the eighteenth- and early-nineteenth-century hospital as monetary exchange was alien to it. Class and dependence as much as diagnosis determined one's place in a "system" of health care sited largely in the home—and in which institutional care was limited essentially to the urban poor.[7] In fact, the late-twentieth-century term *health-care system,* with its assumption of a complex, multilayered, bureaucratic, interactive—and by implication public—world of medicine, is irrelevant to an era without specialists and without laboratories, an era in which the great majority of medical care was performed in the patient's home, whether by family members or professional physicians. The worthy poor were presumed to deserve voluntary hospital care without incurring the stigma that came with almshouse admission. Physicians were presumed to have an obligation to provide gratuitous or discounted care to those unable to afford their fees. Whether rural or urban, nineteenth-century Americans were, that is, presumed to have a right to such care, but not of course to equal—class-blind—care.

The public sector did play a role in the provision of health care, but only in regard to the dependent, not to those seen as able to care for themselves. A socially constructed sense of stewardship, of categorical moral obligation motivated and shaped the efforts of our earliest hospitals' founders. They did not expect to be judged primarily by the success or failure of marketplace decisions (though they were expected to function responsibly and effectively within the market). The medical profession was presumed—at least in theory—to be motivated by a code of gentlemanly and selfless benevolence; patenting discoveries—like advertising one's clinical services—were, for example, seen as evidence of sordid quackery, not rational market behavior.

In 1800 medical ideas and medical practice were widely distributed throughout society—in patterns vastly different from those Americans became accustomed to in the late twentieth century. Conventional moral values suffused both lay and professional ideas of disease causation and treatment, for example, but were not legitimated in terms of modern notions of specific, mechanism-defined disease. Disease categories did not, logically enough, play as prominent a role in lay understandings of behavioral deviance or in physicians' understanding of appropriate therapeutic and diagnostic choices. Homosexual behavior was a willed act of immorality, for example, not a disease, personality type, or merely one among a variety of lifestyle patterns; disruptive grammar school children were wicked and undisciplined, not victims of attention deficit hyperactivity disorder. Death involved prognosis and pain, confrontation with a patient's spiritual and aggregate physiological status—not the management of machines and the hegemony of bureaucratic protocols and insurance schemes. Euthanasia meant literally that—a good death—and implied the deployment of opiates, moral reflection, and family, not respirators and advance directives.[8] Research had not yet been clothed with a transcendence rivaling that of traditional religion and community obligation.

The market too has in some circles acquired its own quality of transcendence, neutralizing traditional suspicions that short-term profit maximization would bring excessive competition. In earlier eras competition was understood not as guarantor of economically efficient and therapeutically efficacious health care, but as an insidious motivation for misrepresentation and shoddy practice.

There are, of course, continuities as well as contrasts between the end of the eighteenth and the end of the twentieth centuries. Chronic disease, for example, posed questions of behavior, volition, and regimen—just as today's anxieties about risk factors and lifestyle mobilize feelings of guilt and accountability.[9] And men and women felt pain, feared death, mourned the loss of loved ones—as they still do.

My argument will have become clear enough by now. I have tried to illustrate in concrete terms the way in which morality and moralism, obligation and responsibility are unavoidably elements of medical care—and at the same time contingent and historical. Medical ideas and practices have always reflected, incorporated, and sanctioned prevailing notions of value and responsibility. Such ethical assumptions imply priorities and constrain choice; meaning and morality are thus necessarily and inextricably embedded in every aspect of medical practice, private and public, individual and collective.

NOVEL REALITIES

If anything can be said to characterize our particular moment in the relationships among the linked histories of medicine, culture, and public policy, it is—as I have emphasized—a novel sense of change and conflict, an uncomfortable awareness of the difficulties inherent in balancing the sacred and the technical, the individual and the collective, of configuring the rights of physicians, individual patients, and the general good. It was, in fact, out of such perceived conflict that bioethics itself developed as a self-conscious movement in the 1960s and early 1970s. Its very creation was in part a symptom—as well as recognition—of perceived inequity, of a gap between medicine's presumably sacred and humane tradition and a reality often egregiously inconsistent. It was an acknowledgement that something needed to be done.[10]

In another sense, this gap between medicine's humane tradition and a more complex and compromising reality can be thought of as a structured crisis in supply and demand: a demand constituted by pain and anxiety and the inexorable realities of demography and chronic disease yet routinely construed in terms of procedures and specialists.[11] Americans have produced a reservoir of insatiable clinical demand ill suited to a world of supply dominated by technology, by impersonal—and costly—providers and products.

This asymmetry embodies a structured conflict that a minority of far-sighted social scientists and physicians has warned about since the Progressive era at the beginning of the twentieth century—when such critics deplored a growing medical impersonality and dependence on what they already saw as an increasingly pervasive technology. Such anxieties might, in fact, be seen as precursors of the late-twentieth-century bioethics movement—an affirmation of the individual and the idiosyncratic as opposed to the depersonalization and fragmentation of care implied by clinical pathology, specialism, and reductionist understandings of health and disease. We have—that is—experienced a century of recurrent crisis in how we think about medicine and what we expect from it. We seem to have created a system in which boundless material expectations are bound to disappoint, and in which we increasingly and paradoxically keep trying to reach personal, that is intangible and experiential—holistic—ends, through technical and mechanism-oriented—reductionist—means.

Let me cite another bit of media evidence to illustrate this point more concretely. A *Newsweek* feature article on the genetic causation not only of clinically well-defined mental illness, but of a bewildering variety of human peculiarities, all construed as less severe manifestations (shadow ailments caused by the presence of one or more "abnormal" genes) of a multigenic illness. "Idiosyncratic behaviors and personality quirks once thought merely 'odd' or 'interesting' might be, in a sense, mental illnesses." The *Newsweek* reporter explained her findings as "a reflection of an abnormality in the brain, and even in the genes."[12]

Though perhaps at first thought unrelated to the previously mentioned changes in such historically significant institutions as Montefiore and the Pennsylvania Hospital or to understanding the social place of bioethics, this story illustrates a fundamental and in fact logically related aspect of twentieth-century medicine: its characteristic search for mechanism-based understandings of an ever wider range of human behaviors. This relentless medicalization of both normal and deviant behavior sheds a parallel and supplementary light on a fundamental structural reality in America's health-care system: the tendency to ask medicine to do more and more cultural work while demanding that this cultural work be legitimated in terms of biological mechanism. It is in part a crisis—as illustrated in the *Newsweek* story on the genetic determination of practically everything— in how we legitimate norms, manage deviance, think about ourselves. Behavior, agency, culture itself can be ingenuously reduced to neurochemical mechanisms—even if this determinism continues to dismay those anxious to maintain a place for human agency and individual responsibility.

This structure of linked ideas and institutional relationships poses a number of problems for both historian and bioethicist. Perhaps most fundamental is the way in which ideas, values, and expectations become embedded in institutions, in practices, in economic relationships and interests. Second is the way in which the concepts and practices of medicine have become increasingly central to the everyday lives of men and women, metastasizing onto the business and editorial as well as the news pages; we seem well on the way to medicalizing not just deviance but also almost every aspect of daily life. Third is the way in which medicine is simultaneously within and outside the market—a paradox that frames today's most vexing organizational question: Can the market (as mediated through public advocacy and the political process) prove adequate as a means of distributing clinical equities and outputs when demand is defined in more than material terms? Can the market produce rational and rationalized—collective—solutions that must be experienced in moral and emotional—individual—terms?

The bottom line as I have tried to emphasize is that we cannot remove or isolate value assumptions from the institutional, the technical and conceptual in medicine; men and women inevitably express their sense of need and priority in the public sphere. Medicine is negotiated and inevitably political and—as we have come to understand more generally—the political cultural. The heated contemporary debate surrounding managed care illustrates in a very concrete way the nature of such interconnections between values and interests. Questions that can be framed as matters of justice and autonomy are at once questions of control and economic gain. Perceptions of right and wrong, of appropriate standards of practice, constitute de facto political realities—variables in negotiating choices among rival policies as well as in particular clinical interactions. The widespread assumption, for example, that it is right for government to play a role in providing and regulating health care is a specific historical and ethical—and thus political—reality, and so is the equally pervasive assumption that it is somehow immoral for mere economic calculation to constrain a physician's clinical decision making. The willingness, in fact, to nurture bioethics constitutes similarly a public recognition of medicine's special moral identity.

But this vague moral consensus cannot mandate a precise and unambiguous social agenda for bioethics. This new enterprise has been charged with a difficult and elusive job. We live in a fragmented yet interconnected world, a world of ideological and social diversity, of inconsistency and inequity, of change and inertia. We cannot discuss relationships among men and women who differ in power and knowledge without acknowledging those inequities: class, geography, gender, race, and education all modify the category *patient;* economic incentives, as well as the institutional and intellectual structures of medicine (such as specialty and organizational affiliation), modify the category *physician.* A growing awareness of such complexities has made bioethics an increasingly labile and self-conscious enterprise. And perhaps a less self-confident one as well; articulating and applying a foundational ethical basis for particular social actions no longer seems an easily attainable goal.

Inconsistent ideas as well as social diversity shape available choices for both physician and patient. American society has elaborated and internalized not a unified and coherent moral consensus—but rather a world of medical discourse and practice marked by the claims of three competing and not always consistent transcendences. One is the academic research tradition with its worship of the selfless search for knowledge—and a widespread faith in its inevitable application. It is a kind of secular millennialism, powerful not simply because it is a source of undifferentiated cultural optimism, but because it is structured into the expectations and hopes of individuals—into the career choices of particular physicians and scientists, into the formation of public policy, and into the status and programs of academic departments and teaching hospitals.[13] Second, and more recent, is the worship of system as goal and ideal, of the assumption that the optimum general good is attainable only through an optimum configuration of market and institutional relationships. Finally, of course, is the traditional moral specialness of medicine, of respect for physician responsibility and the rights of individual patients—a tradition that can be traced from classical antiquity to contemporary debates over medical care. Each of these claims to transcendence legitimates claims to social authority; all are ceaselessly configured and reconfigured as medicine's technical resources and institutional forms evolve and pose novel research and clinical options. Bioethics has, in fact, already become a useful supporting actor in the complex interactions that characterize relations among these realms of value and implicit power.

I have tried in the preceding pages to illustrate a number of the ways in which the moral values that suffuse medicine are historically constructed and situationally negotiated—like every other aspect of culture—and not simply derived from the formal modes of analysis that have historically characterized theology and moral philosophy (though such delineations of fundamental principle are in themselves an element in the social negotiations that inform and rationalize health care). The formulations of credentialed philosophers and theologians are at once a claim to cultural authority and a factor in the public mediation of social conflict.

The very existence of a socially visible enterprise called "bioethics" is a recognition of the recurrent structured conflicts I have tried to illustrate anecdotally. Thus, I began this discussion with particular examples of institutional change—and my construction of them as both morally and policy relevant. In some ways, of course, these are overlapping categories; moral certainty *is* a political and policy reality. I wanted to emphasize the ways in which its history underlines medicine's context dependence and, in particular, of the way in which medicine necessarily embodies a variety of attitudinal and value elements as well as technical capacity and institutional practice.

But this is only one of the ways in which bioethics and history relate. First, from the historian's disciplinary perspective, bioethics is a complex and potentially revealing subject for empirical investigation. Second, and more important, I would contend that although academic history and bioethics have in general followed separate paths, they share a potential community of sensibility, a sensitivity to context and to the relationships among individual perception, social constraint, and the situatedness of human agency. Practitioners of history and bioethics should, finally, be similarly aware of the importance of irony and contingency, of the gap between theory and practice, conscious intent and unforeseeable outcome.

The still brief history of American bioethics demonstrates just such realities. As a social movement, bioethics developed in mid-twentieth century as a critical enterprise—a response to felt inhumanities in our system of health care and biomedical research. A response to specific abuses, bioethics has remained practice oriented; society expects bioethics to solve or at least ameliorate insistently visible problems.

Growing as it has out of a sense of moral outrage, bioethics has had an undeniable impact on everyday clinical realities. Yet, from the historian's perspective, this novel enterprise has played a complex and in some ways ambiguous role. Bioethics has not only questioned authority, it has in the past quarter century helped constitute and legitimate it. As a condition of its acceptance, bioethics has taken up residence in the belly of the medical whale; although thinking of itself as still autonomous, the bioethical enterprise has developed a complex and symbiotic relationship with this formidable host organism. Bioethics is no longer—if it ever was—

a free-floating, oppositional, and socially critical reform movement: it is embodied in chairs and centers, in an abundant technical literature, in institutional review boards and consent forms, in presidential commissions and research protocols. It can—that is—be seen as a mediating element in a complex and highly bureaucratic system that must, nevertheless, manage ceaseless technical change. It is not an accident that the bioethical enterprise has routinely linked bureaucracy—committees, institutional regulations, and finely tuned language—with claims to moral stature.

But this functional role implies a structured conflict. By invoking and representing medicine's humane and benevolent, even sacred cultural identity, bioethics serves ironically to moderate—and thus manage and perpetuate—a system often in conflict with that idealized identity. In this sense, principled criticism of the health-care system serves the end of system maintenance. It is such paradoxes of power and consciousness that explain why bioethics needs to think of itself both historically and politically, and in some ways this process has already begun.[14] This novel enterprise has already enshrined its heroes and villains—Henry Beecher and Josef Mengele—and commemorated its sacred places—Willowbrook, Tuskegee, Nuremberg. In fact, one could argue that the historical stocktaking initiated by bioethics' founding generation is itself an aspect of what might be called institutional consolidation.[15]

Participant histories serve celebratory and mystifying as well as analytical and self-critical ends. History can be used as source of both false-consciousness and the celebration of conscience. It is difficult for the committed practitioner not to foreground her field's positive values and accomplishments, not to see herself on the side of the angels, fighting the good fight against the routine and unself-conscious abuse of men and women in everyday clinical and research settings. It is equally difficult to see the apparatus of committees and regulations that protect patient rights against the abuses of an impersonal technology as itself a technology. By way of example, let me quote the words of a bioethicist reacting to an earlier version of my present remarks—and in particular to a passage in which I described the bioethical enterprise as in some ways a technology necessarily mirroring the technology it sought to meliorate. "Bioethics," the indignant reader explained, "in the late twentieth century in American medicine has always championed the rights of the individual patient against the vagaries of the medical system. Its cardinal principles of autonomy, beneficence, non-maleficence, and justice represent the antithesis of technology."[16]

Most contemporaries would not be quite so uncritical in their self-evaluation—yet they are still ill prepared to deal with what I have characterized as the central irony of bioethical success: insofar as it has been accepted by the world of research and clinical practice, it has become a part of those linked enterprises—and thus its every criticism and consequent procedural reform cannot help but constitute an aspect of biomedicine's public moral face.

As a specific empirical subject, moreover, bioethics presents an elusive aspect—as elusive as weighing its ultimate social impact. In part this is because the bioethical enterprise is an aggregate of three not always consistent activities. One is the elaboration of formal doctrine—the job of individuals trained to articulate and address normative ethical questions. I refer, of course, to those philosophers and theologians who have sought to create a principled consensus around such policy-defining issues as autonomy, beneficence, and justice. Second is the role of bioethics in mediating day-to-day clinical problems in particular social settings. I have in mind the innumerable contexts in which institutional review boards, the deliberations of government commissions, the language and ritual of informed consent, make practitioners and researchers aware of the rights of patients and subjects. Third is the way in which the bioethical enterprise figures in public discourse, responding in newspapers, periodicals, television, and—in recent years—the Internet to novel dilemmas derived often but not always from technological innovation. Fourth—and closely related to the third—bioethics serves as a factor in the formulation, rationalization, and legitimation of public policy. In this highly visible capacity bioethics reassures, implying that there is a discernible moral order that can be used rationally to manage new and potentially alarming clinical and research choices. It is both ritual and spectacle, acting out the several reassurances of ethical concern, credentialed expertise, and the assumption that fundamental ethical principles can be discerned and applied.

Thus, bioethics occupies three distinct (if often overlapping) social spaces. One is academic—formal, discipline and text oriented. A second is the institutionalized presence of bioethics in hospital and research settings. Third, as I have suggested, is the media. This mosaic of roles and sites of social action makes bioethics both complex in structure and difficult to delineate. This diversity of site, personnel, and function also explains my avoidance of the term "discipline" in describing what I have chosen to call instead the "bioethical enterprise"—a conglomerate of experts, of practices, of ritualized and critical discourse in both academic and public space.

BIOETHICS AND THE HISTORICAL SENSIBILITY

I have specified a number of ways in which bioethics and history might share an analytic perspective. First, and perhaps most fundamentally, I would argue, the task of ethical understanding should parallel the historian's job of cultural reconstruction: both kinds of practitioner should seek—if necessarily imperfectly—to understand a time- and place-specific structure of choices as perceived by particular actors. Second, I would argue that we cannot understand the structure of medical choice without an understanding of the specific histories of medicine and society that have created those choices. This was the argument I hoped to illustrate in my earlier references to change in contemporary American hospitals and my emphasis on increasingly reductionist understandings of disease. And third, and perhaps most disquieting, we must historicize bioethics itself. For it is clearly a place- and time-bound enterprise, with complex relationships to the special world of medicine and to the larger society in which medicine is nurtured and which medicine in part constitutes.

My first point, which seems no more than a truism to a cultural historian, will seem irrelevant or perhaps even philistine to scholars focused on the elucidation of ethical principles abstracted from precise social and institutional contexts—even if they are motivated by abuses at just such specific sites. Moreover, such formal styles of normative discourse have paralleled and intensified the historical tradition of medical ethics with its emphasis on the unmediated physician-patient dyad—one physician, one patient, one bedside, the paradigmatic vexed case. From the contextually oriented historian's point of view, however, choice is always constrained and structured, a reality to be understood in specific situations, not schematically in terms of logically and morally coherent ends. In this historical and sociological sense, autonomy is a product not a goal; it is a place-, time-, and system-specific outcome of the interaction between the microcosm of the clinical encounter and the macrocosm(s) of the larger society and the cognitive and institutional world of medicine. This needs hardly to be elaborated at a moment in time when many physicians find their clinical interactions limited by managed-care providers to fifteen minutes and their diagnostic and therapeutic choices limited as well. Autonomy and agency are constructed and reconstructed in every healing context. There can be no decontextualized understanding of bioethical dilemmas; bioethics is definitionally contextual, as I have argued, finding its origins in the search for particular solutions to visible social problems. A decontextualized approach in bioethics is not simply a matter of disciplinary style; it is a political act.

Discussions of informed consent, for example, that abstract the actors—clinicians, researchers, patients, and "subjects"—from their particular social roles and individual identities are not very helpful—and must, in fact, mystify these social relationships and, in doing so, legitimate the de facto authority of those individuals and institutions doing the "consenting."[17] At the risk of seeming didactic, let me take a moment to underline the way in which the colloquial use of "consent" as a verb illuminates the ambiguity of routinization in the management of "autonomy" and "beneficence." This usage is a syntactical representation of power and comparative powerlessness, of actor and the object of that actor's actions. To consent a patient is to act out—and legitimate—a reality of social inequality as well as to demonstrate the existence of a self-conscious community of "consenters" well aware of the ritual and hierarchical aspect of this now pervasive—and paternalistic—ethical mechanism.

I would argue, moreover, that bioethics is not only defined by its context of use but that it cannot be self-aware without an understanding of the history of medicine in the past century—of the role played by new and specific notions of disease, by the growth of specialism and credentialing, by the siting of the clinical encounter in a technologically rationalized and structured institution instead of the patient's home or physician's office. This point hardly needs elaboration. Bioethics is or should be a social, an ethnographic, and a historical enterprise, for the issues it seeks to mediate are themselves the products of a specific, determining history. Without history, ethnography, and politics, bioethics cannot situate the moral dilemmas it chooses to elucidate. It becomes a self-absorbed technology, mirroring and inevitably legitimating that self-absorbed and all-consuming technology it seeks to order and understand.

But, as I have suggested, it is easier to call programmatically for bioethics to place itself and its tasks historically than to accomplish that elusive task.

There is no simple path to understanding the historical place of bioethics, but rather a variety of interpretive options, reflecting the interpreter's point of view and the inherent elusiveness of the subject. How one construes the enterprise reflects a diversity of interpretive perspectives. To some critics on the left, bioethics is no more than a kind of hegemonic graphite sprayed into the relentless gears of bureaucratic medicine so as to quiet the offending sounds of human pain. Its ethical positions—this argument follows—are in terms of social function no more than a way of allaying social and legal criticism, and the self-reproaches of a minority of ethics-oriented physicians. It has, moreover, this critical position contends, focused too narrowly on the visible problematic instance—on the plug pulled or not pulled, on the organism cloned or the cloning interdicted—and avoided consideration of less easily dramatized policy debates and mundane bedside dilemmas. And, finally, the argument follows, it is not surprising that in a bureaucratic society we have created a cadre of experts and body of knowledge to provide a soothing measure of certified and routinized humanity.

To its sophisticated practitioners and advocates, on the other hand, bioethics is a humane change agent, an important mechanism for mediating technological and institutional change—a kind of software that facilitates the adaptation of novel varieties of hardware. It is, the argument follows, a genuine constraint, a substantive actor in a complex renegotiation of everyday medical practice; bioethics has—similarly—influenced the conduct of clinical research with human—and animal—subjects. One need only point to the creation of research guidelines for human and animal subjects, to the existence of institutional review boards, and to good-faith attempts to make informed consent a reality. Even if an unfettered individual autonomy may be an unrealizable ideal, the assumption nevertheless that there is such a thing contributes to a viable framework for thinking about transcendent value, constitutes in itself a resource in the complex negotiations that determine and constrain individual and institutional choice. Bioethics has also played a constructive role in the public discourse surrounding clinical medicine and biomedical innovation, a media discourse that is necessarily focused on particular problems as spectacle, yet such perception-altering public discourse helps redefine the structure of real-world political options.[18] Perhaps most important, bioethics expresses the widely felt social—and thus political—assumption that medicine is and must be more than a sum of technical procedures and market transactions.

It promises solutions to human dilemmas beyond the profit-maximizing choices of the market or the ultimately elusive if seductive dreams of technological utopianism.

HISTORY, CONTINGENCY, AND BIOETHICS

If the three principles of value in real estate are—in the language of an old saying—location, location, location, history's are context, context, context. And irony and contingency are implicit in a contextual style of analysis; history, like life itself, is filled with unintended consequences. In one significant respect, though, historians are more fortunate than bioethicists: no one expects them to solve emergent social problems. The bioethical enterprise, on the other hand, originated—as we have seen—as a response to such perceived problems and continues to offer not just analysis but solutions to them.

Yet the most profound of such problems are in their nature unsolvable. We are well aware that there is no ultimate solution for pain and death, no way to explain the brutal randomness with which suffering is distributed. These are aspects of the human condition. Some other issues are perhaps less obvious. There is also no easy solution, for example, for the way inequalities of social identity reenact themselves in medical care. Moreover, health, disease, and medical care mobilize deeply internalized cultural and religious values and in few societies are such values distributed uniformly; conflict is almost inevitable. Another paradox grows out of our natural, yet contradictory, desire for cure and care, for technological efficacy with a human face. But care and cure are not easily linked in the same context; the historical circumstances that produce the laboratory's undeniable achievements also produce the bureaucracy that intimidates, fragments, and distances. A parallel conflict grows out of the difference between interest as defined by the individual and by the collective; a test or procedure that can benefit one individual might be irrational from the social-system perspective. It is a system, moreover, that has consistently demonstrated the ability to incorporate the critical and morally oppositional and make it an aspect of the system itself. And thus, perhaps, the ultimate irony of bioethics' history—the persistent, yet perhaps illusory quality, of our desire to routinize the humane, to formulate and safeguard timeless values in worlds of ceaseless change, social inequality, and utopian laboratory expectations.

NOTES

1. Esther B. Fein, "Region to Get Clinics Giving Specialty Care," *New York Times,* February 9, 1998, B1.

2. Andrea Gerlin, "A Venerable Phila. Hospital Charts a New Course," *Philadelphia Inquirer,* February 1, 1998, D1. For background on the sale of the Institute of the Pennsylvania Hospital, see Karl Stark, "Talk Isn't Cheap, Forcing Changes in Psychiatry," *Philadelphia Inquirer,* September 14, 1997, E1.

3. Even more recently, the Philadelphia region's health-care system has been destabilized and demoralized by the aggressive takeover strategy of the Allegheny Health, Education, and Research Foundation, a Pittsburgh-based health care system, which purchased physician practices, hospitals, and associated medical schools in a bold marketplace venture that soon ended in bankruptcy, unmet commitments, and a perilous future for such historically significant institutions as Hahnemann Medical College and the Medical College of Pennsylvania. Hahnemann was founded as America's premier homeopathic training institution in 1848, and the Medical College of Pennsylvania is better known to historians by its original appellation: the Woman's Medical College, a pioneer institution founded in 1850. In 1993, the two schools announced their intention to merge under the auspices of the Allegheny Health, Education, and Research Foundation; they admitted their first class two years later. See Lawton R. Burns and Alexandra P. Burns, "Policy Implications of Hospital System Failures," in *History and Health Policy in the United States: Putting the Past Back In,* ed. Rosemary Stevens, Charles E. Rosenberg, and Lawton R. Burns (New Brunswick, N.J.: Rutgers University Press, 2006), 273–308; Naomi Rogers, *An Alternative Path: The Making and Remaking of Hahnemann Medical College and Hospital of Philadelphia* (New Brunswick, N.J.: Rutgers University Press, 1998); Gulielma F. Alsop, *History of the Woman's Medical College: Philadelphia, Pennsylvania 1850–1950* (Philadelphia: J. B. Lippincott, 1950); Steven Peitzman, *A New and Untrue Course: Woman's Medical College and Medical College of Pennsylvania, 1850–1998* (New Brunswick, N.J.: Rutgers University Press, 2000).

4. Robert Pear, "Medical Research to Get More Money from Government: An Investment in Health," *New York Times,* January 3, 1998, A1.

5. Pam Belluck, "Heartache Frequently Visits Parents with Multiple Births."

6. *Time,* January 11, 1999.

7. Which is not to say that such variables are not significant. See, for example, Charles E. Rosenberg, *The Care of Strangers: The Rise of America's Hospital System* (New York: Basic Books, 1987); Morris J. Vogel, *The Invention of the Modern Hospital: Boston, 1870–1930* (Chicago: University of Chicago Press, 1980); and Dorothy Levenson, *Montefiore: The Hospital as Social Instrument, 1884–1984* (New York: Farrar, Straus & Giroux, 1984).

8. See, for example, William Munk, *Euthanasia; Or, Medical Treatment in Aid of an Easy Death* (London: Longmans, Green, and Co., 1887).

9. See Allan M. Brandt and Paul Rozin, eds., *Morality and Health* (New York: Routledge, 1997), and "Banishing Risk," chapter 4 in this volume.

10. Although it is conventional to see the bioethics movement as having its late-twentieth-century origins in Nuremberg, it is equally conventional to see it crystalliz-ing as a self-conscious movement in the 1960s as in part a response to those social currents that produced a more general sensitivity to individual rights—of women, of prisoners, of sexual and racial minorities—and of an antiauthoritarian skepticism to-ward credentialed expertise. For a first generation of bioethics history, see Albert R. Jonsen, *The Birth of Bioethics* (New York: Oxford University Press, 1998); David J. Roth-man, *Strangers at the Bedside: A History of How Law and Bioethics Transformed Medical Decision Making* (New York: Basic Books, 1991); George J. Annas and Michael A. Grodin, eds., *The Nazi Doctors and the Nuremberg Code: Human Rights in Human Experimentation* (New York: Oxford University Press, 1992); and M. L. Tina Stevens, *Bioethics in Amer-ica: Origins and Cultural Politics* (Baltimore: Johns Hopkins University Press, 2000).

11. I do not mean to imply that medical practice has, in fact, been consistently humane, caring, and selfless over time—but rather that a commitment to this ideal has always been part of the profession's formal corporate identity and structured into the physician's sense of self and social legitimacy.

12. Sharon Begley, "Is Everybody Crazy?" *Newsweek,* January 26, 1998, 51–55.

13. For an argument emphasizing the nineteenth-century roots of such disci-pline-structured patterns of value and action, see Charles E. Rosenberg, *No Other Gods: On Science and American Social Thought,* 2nd ed., rev. (Baltimore: Johns Hop-kins University Press, 1997).

14. See Jonson, *The Birth of Bioethics;* Rothman, *Strangers at the Bedside;* George Weisz, ed., *Social Science Perspectives on Medical Ethics* (Philadelphia: University of Pennsylvania Press, 1990); Albert R. Jonsen, ed., "The Birth of Bioethics," special supplement, *Hastings Center Report* 23 (1993); Ruth R. Faden and Tom L. Beauchamp, in collaboration with Nancy M. King, *A History and Theory of Informed Consent* (New York: Oxford University Press, 1986).

15. For a deeply informed guide to this history by an influential participant, see Jonsen, *The Birth of Bioethics.* Particularly relevant to the present discussion are Jonsen's chapters 10 and 11: "Bioethics as a Discipline" and "Bioethics as a Dis-course," 325–76.

16. Personal communication, Sheldon Lisker, M.D., to the author, February 1, 1999.

17. For an illuminating case study, see Renée R. Anspach, *Deciding Who Lives: Fate-ful Choices in the Intensive-Care Nursery* (Berkeley: University of California Press, 1993).

18. The recent media achievement of Dr. Jack Kevorkian in making end-of-life issues a public issue is a case in point. It is not surprising that formative episodes in the history of American bioethics have been associated with dramatically visible events and individuals—from Karen Ann Quinlan to Dolly the cloned sheep.

- -

CHAPTER 10

ANTICIPATED CONSEQUENCES

Historians, History, and Health Policy

"POLICY" IS A FAMILIAR TERM, BUT, LIKE MANY INDISPENSABLE WORDS, it is not easily defined. In one sense it is descriptive: policy refers to current practice in the public sector. It also has a variety of other meanings: policy may imply an "ought" of planning and strategic coherence—or a real world "is" of conflict, negotiation, and compromise.[1]

As the history of United States health policy makes clear, moreover, the real world is not a very orderly place. Policies on the ground seem less a coherent package of ideas and logically related practices than a layered conglomerate of stalemated battles, ad hoc alliances, and ideological gradients, more a cumulative sediment of negotiated cease-fires among powerful stakeholders than a self-conscious commitment to data-sanctioned goals. But policy outcomes are hardly random; they embody the divergent rationalities and strategies of contending interests. Public sector outcomes are determined by structured contention and contingency—not the prospective models and metrics of social scientists.[2]

Thus, the familiar dismissal of the historical community's potential contribution to policy seems, at least to this historian, paradoxical. Structured contention and contingency *are* history, and so is contemporary policy—even if historians and historical data seem tangential to the demanding (and demanded) task of anticipating the consequences of particular present actions.[3] From the historian's perspective, it is equally clear that recent policy—including ideological invocations of the past—constitutes a comparatively neglected area for research. Health policy tells us a great deal about the relationships among interest and ideology, formal structures and human need, professionalization and social welfare, and technology and its applications. The debates, rationale, and actions of successive generations of American legislators, executives, credentialed experts, and administrators have constituted a sequence of what might be described as recurring social experiments and thus rich data for the historian and policymaker.[4]

Some of the key themes in this particular story have become abundantly clear. One is the foundational relationship between public and private functions, peculiarly significant in medicine but far from irrelevant in other areas of American life. Second is the way in which values and ideas shape perceptions of the possible and the desirable and express themselves in the social strategies of institutions, disciplines, and interest groups. Third is the way that the policy-making process constantly creates new realities and new choices, yet is itself structured by preexisting interests, perceptions, and cumulative decisions. Finally, I would emphasize the persistence of structured—and thus to a degree predictable—conflicts shaping American health care, conflicts that have grown, and will continue to grow, out of interactions and inconsistencies among these three underlying factors.

All of these themes are fundamental to not only health policy but also American history more generally. Medicine is an indicator as well as a substantive component of any society. It is a cliché to emphasize that all medicine is social medicine and that the provision of health care links individual lives to larger social and cultural realities—but this cliché has not always dictated academic research priorities. The seemingly meandering history of American health policy underlines this instructive, if banal, truth.

PUBLIC AND PRIVATE

There is nothing more fundamental in the history of American health care than the mixture of public and private. In this regard, American distinctiveness lies not in some unique devotion to the market and individualism, but in a widespread inattention to a more complex reality. From the canal and railroad land grants in antebellum America to support for the aircraft industry in the twentieth century, from tariff policy to the creation of the corporation in the nineteenth century to today's outsourcing of military functions, the interactive and mutually constitutive mixture of public and private has been so ubiquitous in American history as to be almost invisible; it is as true for medicine as it is and has been for transportation or the military-industrial complex. All have been clothed with a sense of collective responsibility that implies—if not demands— the active role of government. Since World War II, the public sector (and especially the federal government) has supported medicine in all of its aspects—basic research and the training of biomedical scientists and clinicians, the provision of care, and the management of medically defined dependency.

It is a tradition with roots older than the nation itself. The Pennsylvania Hospital, for example, was founded in the early 1750s through a mixture of legislative subvention and private philanthropic money and leadership. Similarly, in colonial Virginia, the House of Burgesses created the colonies' first mental hospital to deal with a community responsibility, the dependent insane. Dependency and sickness have always enjoyed a complex and symbiotic relationship in America. Almshouses, for example, always had a medical-care component, even if one not easily distinguishable from their general welfare function.[5] Nineteenth-century localities found numerous ways to support health care—from lotteries to tax policy to cash grants—while states created institutions to care for the insane, the tubercular, and in some states later in the century even the epileptic and the inebriate.

It seemed only natural. Health and medicine have always been seen as clothed with the public interest, as somehow different from—if dependent upon—economic relationships; medicine was in, but not entirely of, the market. And if pain, need, and sickness have made medicine historically sacred, the relationship between sickness and dependency in another of medicine's aspects has linked the provision of treatment and care to the state's welfare responsibilities as well as to traditional notions of religious benevolence and more contemporary assumptions concerning access to health care as an aspect of citizenship in a democratic society. This unself-conscious yet intricately interdependent relationship between the public and the private, the community and the individual, the spiritual and the material, has always characterized health care in America. Public and private are ultimately no more distinguishable in medicine than are art and science or care and cure. In the area of health, government and the market, public and private are hard to imagine as distinct and exclusive domains, except as ideological—if analytically useful—abstractions; in the real world they cannot be understood in isolation from one another.

It is a historian's task to understand the structure of such interrelationships. How, for example, are we to think about the creation, the regulation, and the clinical use of pharmaceuticals? Despite an intense contemporary focus on the decisions of corporate actors, the reality is far more complex. Government pays for much of the basic research and the training of laboratory investigators at every level and has for a century played a regulatory role, and more recently acted in the market as purchaser (directly or indirectly) through Medicare, Medicaid, and the Veterans' Administration. Patent law and the courts constitute another significant aspect of government involvement in the world of medical therapeutics. The medical profession also plays a fundamental role, serving as a source of intellectual and moral authority in the acceptance and clinical administration of drugs—thus occupying a position at once private yet clothed with public interest and authority.[6] It is within this context that we must view the vexed contemporary question of payment for prescription drugs, an increasingly visible, intractable, political—and politicized—issue that links the private and the public, the corporation and the community.

This pattern of what might be called mixed enterprise in medicine is hardly limited to the worlds of drugs and their clinical use. The institutional history of American medicine reflects the same pattern; hospitals and outpatient services have always been clothed with the public interest, for example, yet have remained largely in the hands of nongovernmental—not-for-profit—entities. It is assumed that hospitals perform basic social functions in restoring the sick to productive social roles—as well as serving a higher, less instrumental and more spiritual, good. As I have just suggested, even the regulation of the medical profession reflects the same pattern. Specialty boards are private entities, yet they assume de facto public responsibilities, and their policies and credentialing activities create social, legal, educational, and even economic realities.

I have argued that for a mixture of instrumental and moral reasons it has long been assumed that the state has some role—and an interest—in protecting the health of its citizens generally and of providing at least minimal care for the helpless and indigent. This generalization is hardly controversial, but insufficiently precise. Cultural assumptions, institutional forms, and technical capacity are historically specific—as is medicine itself. Until the mid-twentieth century, few in the Anglo-American world thought it the responsibility of the public sector to support medical research, and not until the end of the nineteenth century did most laypeople think it important to control access to the medical market by carefully designating legitimate practitioners.[7] But even in America's earliest years as a nation, it was assumed that communities might and should enforce quarantines and a minimum level of urban sanitation as well as provide some care for the chronically ill and incapacitated. (As early as the end of the eighteenth century, for example, the federal government initiated a health insurance scheme to protect merchant seamen, workers regarded as strategically important and peculiarly at risk.) Obviously our social assumptions have changed drastically since the Federalist era—and the role of the state in medicine has been revolutionized, particularly in the last three-quarters of a century.[8] Nevertheless, this theme of a role for the public sector in protecting community and individual health has remained a reality, even if continually redefined, renegotiated, and continuously—if erratically—expanded. And this despite a quarter century of rhetorical demands for smaller government; the rhetoric has remained just that—words, even if politically resonant words—as the public sector's role in health care has hardly diminished.

A parallel and reinforcing ambiguity surrounds the medical profession and its relationship to the market. Medicine has always been a business, and American physicians have until comparatively recently, in historical terms, had to earn their living in a brutally competitive and unforgiving search for paying patients. The professional identity and market plausibility of medicine, however, have rested historically on a special moral and intellectual style, formally transcending the material reality and reflecting the sacredness of human life and the emotional centrality and specialness of the physician-patient relationship. In the past century, the medical profession's claims to self-regulation and autonomy of action have been justified by their mastery of an increasingly efficacious body of clinical knowledge. Physicians are no longer priests, as they sometimes were in the premodern world, and many other contemporary professions claim a legitimacy based on esoteric knowledge. Nevertheless, there remains something special about the physician's vocation, about the profession's peculiar configuration of ethical and knowledge-based claims. Even in those centuries when, as gentlemen, physicians could not charge fees but expected honoraria, predecessor versions of this inconsistent but seemingly functional set of values defined the special place of medicine and helped legitimize its guild demands for status and autonomy. One consulted an elite physician in part because of the practitioner's moral stature and gentlemanly bearing, but one assumed that learning and skill were the natural accompaniments of such attributes.

This mixed tradition of moral standing and technical expertise has, over time, worked to facilitate the mutually reinforcing interconnection of public and private in medicine, even as the scale and scope of such relationships have changed drastically in the past two-thirds of a century. Medicine's traditional identification with the sacred, the selfless, and the public interest has blurred and hybridized with the intellectual, the technical, and the instrumental. The merging of these diverse sources of authority has obscured areas of potential inconsistency and conflict. Some of these conflicts are obvious. Are physicians necessarily and appropriately profit maximizers or committed to a selfless devotion to their fellow citizens? Are they healers or scientists? And how can these inconsistent visions of the profession's social role and responsibilities be made coherent in the categories of public discourse and as components in the setting of health policy? Such questions underline a few of the ways in which public and private values and policy ceaselessly and inevitably interact—and are in a variety of ways mutually constitutive.

VALUES AND STRUCTURES

I have always been dissatisfied with our conventional usage in which the term health-care system refers only to economic and administrative components.[9] System implies interaction and inclusion—and, in the case of health care, a myriad of cultural expectations and norms, of institutional, political, and technical as well as economic factors. This portion of my argument might be called the "cultural politics" of health policy—a necessary parallel to the "political economy" of health policy. The connections and constraints are ubiquitous. One need hardly elaborate when we have in recent years lived through debates on abortion, on cloning and stem cells, on managing the end of life, on the provision of care and drugs to the aged, and on the expectations surrounding basic and applied science. The list could easily be extended.

It is easy enough to specify cultural values particularly relevant to medicine: attitudes toward life and incapacity, toward the old and young, toward technology and death, and toward the bodies of men and women. But health policy has, of course, also been shaped by values and rhetorical strategies reflecting general—that is, not specifically medical—attitudes toward individual responsibility (distinctions among work, poverty, and dependence, between the worthy and unworthy poor), toward government, toward the market as mechanism and optimum allocator of social goods. Similar to these assumptions are the cluster of cultural values and practical expectations placed on technical solutions (or temporary fixes as critics might sometimes call them) to intractable social and policy questions. Such attitudes have unavoidable political implications; technological innovation, for example, can be presented and often understood as politically neutral—and thus more easily funded and adopted. The costs and benefits become apparent only in retrospect.

Most of us harbor a pervasive faith in the world of scientific medicine, a visceral and inarticulate hope of a temporal salvation: modern medicine will extend life, avoid pain, provide a gentle death. Even if we regard it as romantic or delusive, the idealization of research and its presumably inevitable practical applications has always been a part of the emotional and institutional reward system of science—and thus of medicine. Such hopes are widely internalized in lay minds as well. One thinks of Sinclair Lewis's *Arrowsmith* (1925) or Paul de Kruif's *Microbe Hunters* (1926), which glorified bacteriologists and public health workers as selfless warriors opposing infectious disease. One thinks also of a newer generation of hopes surrounding the promise of gene therapy and stem cell research.[10] Recent controversy over the commercial exploitation of academic science has in part mirrored this lingering sense of value placed on what used to be called freedom of research—and to the intuitive notion that knowledge of the natural world is the property of all men and women (and certainly the American taxpayer who has paid for a good portion of that biomedical knowledge). Neither assumption fits easily with market-oriented practices that emphasize private sector actors, material incentives, and the blurring of lines between academic and corporate identities.[11]

Another value now widely disseminated is what might be called reductionism in medicine. Most patients as well as physicians expect disease to be a consequence of biopathological mechanisms and thus ultimately understandable and treatable. These views are historically—if perhaps not logically—in conflict with widely disseminated antireductionist ideas such as the interactive relationship between mind and body, the body's natural healing power, the social roots and multicausal nature of disease, and the caring, as well as curing, mission of medicine. Though hardly monolithic in espousing reductionist views, organized medicine generally accepts and rewards the laboratory's achievements and benefits from the status and—often unrealistic—public expectations that accompany technical innovation. There are a good many social consequences of our tendency to see disease as discrete and mechanism-based. Thus, for example, ailments with behavioral manifestations and no agreed-upon mechanism inevitably occupy a kind of informal second-class status in terms of everything from insurance reimbursement to social legitimacy, while—at the other end of the value spectrum—there are unrealistic, or at least premature, expectations surrounding the seductive certainties of genetic medicine.[12]

At the same time, however, Americans have not abandoned a pervasive desire to maximize the role of individual responsibility in the etiology and management of illness. Cultural needs to find a logic of moral accountability in health outcomes inform discussions of any ailment that might be associated even remotely with potentially culpable personal choice. I refer here, for example, to such familiar battlegrounds as the appropriate public health response to AIDS or to smoking or to fetal alcohol syndrome or to substance abuse. It is easy enough to fault the alcoholic, the lung cancer patient, or the depressive for their respective failures of will and thus complicity in their own complaint. Society still needs victims to blame.

As all these examples indicate, historically situated cultural assumptions (or values) both legitimate and structure public policy—not because they float in some unanchored cultural space, but because they are reified and acted out in the policies, interests, self-perceptions, and moral hierarchies of particular men and women and thus in their choices among policy options. Embedded in particular senses of individual self and institutional notions of corporate selves (medicine and nursing, for example), such values shape assumptions and constitute political constraints and motivations.

PROCESS, SYSTEM, STRUCTURE

One of the characteristics of policy is what might be called its cumulative, developmental, or process aspect: each decision and its consequences interact over time to define a new, yet historically structured, reality, often one not anticipated by most contemporary actors. The system moves through visible decision points, elaborated by subsequent administrative practice—with that specific experience along with other relevant variables shaping the next visible shift in public policy. All are linked in a context of periodic confrontation, negotiation, and renegotiation—a setting in which the historian's contextual point of view can be particularly helpful. Policy formulation and subsequent implementation provide us with opportunities to see the relevant costs and benefits as perceived by particular men and women as they choose among a variety of available options at particular moments in time.

But, of course, the health-care system is, from one necessary perspective, a dependent variable; it is part of a political system. In the years since the presidency of Franklin Roosevelt, health has become a substantive and increasingly visible issue in national, state, and even local politics. And I refer to not only the delivery of clinical services but also policies on the environment, income distribution, and lifestyle: all can have effects on health. Increasingly the world of medicine has reflected, embodied, and been subject to the realm of electoral politics and to changes in party structure, in ideology, and in its relationship to other policy issues and the setting of budgetary priorities. Lobbying, local politics, and legislative committee considerations have all played a role in defining the specific contours of health-care legislation—while creating an ever-expanding bureaucracy with inevitable feedback into the relevant service communities and institutions.[13] Every public program creates or reconfigures an interested constituency: hospitals, pharmaceutical and device manufacturers, nurses, and physicians all potentially benefit from or are hurt by changes in government policy; every proposal implies winners and losers. In the largest historical sense, bureaucracy and technology have created a new medicine. And the political system, like the health-care system itself, necessarily incorporates values, perceived equities, available technologies, and institutional forms, which collectively define and constitute the possible and map the desirable.

Though much change is gradual, incremental, and elusive, the nature of policy implies periodic public discussion and decision making. How is health care to be financed? Hospitals to be reimbursed? Dependent children to be cared for? Sexually transmitted diseases to be prevented? How are drugs to be certified as safe and efficacious? The discussions and conflicts surrounding such issues and their resolutions—however provisional—tell us a great deal about political goals and tactics. Sequences of events create a constantly revised sequence of structured choices. By following those events, the historian can focus on the relevant loci of interest, power, and authority as they configure themselves in the consideration, passage, and consequent administration of particular policies. If the health-care system includes values, individual and group interests, and history, the political process allows us to look at that system and, in a sense, weigh those variables. One thinks, for example, of the roles played by medicine and medical specialties, pharmaceutical companies, patient advocacy groups, insurance companies, and party strategists in the recent discussion of prescription drugs. Such analysis demonstrates that the medical profession is not always monolithic, that specialties may have different interests, that academic and community physicians may similarly have different interests, even if all share certain guild interests and attitudes. Nurses, as we are well aware, have interests that differ dramatically from those of physicians, or hospitals. We are equally well aware that for-profit and not-for-profit hospitals may have different interests, as might community and university teaching hospitals—yet all appeal to widespread ideas about technology and clinical efficacy. All also interact with local, state, and national government. The health-care system includes lobbyists and party strategists just as it incorporates cultural norms, hopes rationalized in terms of technology, and the disciplinary identities of physicians and nurses. Nothing ordinarily happens without alliances, and the nature of those alliances may tell us a good deal about underlying structures—as well as the political process. Studying the history of health-care legislation and implementation is a bit like stepping behind the scenes at the proverbial sausage factory; it allows us to study the perhaps less-than-elevating realities of what we delicately call "policy formation." Serious students of health policy are as much obligate political scientists as historians and ethnographers.

Tracking that political process leads us to structure, and analysis of structure permits a finer understanding of sequence, which allows us to configure cultural values and economic and guild interests and even weigh the input of credentialed experts and the data they collect and deploy. One of the policy questions most needing attention is, in fact, the role of would-be rationalizers and problem solvers. How do we understand the impact of those individuals who work in policy-oriented think tanks, then move laterally into the academy, into consulting, or into government? Do they initiate change or simply provide an ensemble of rationales and models—tools in a toolbox—that can be used for particular purposes by strategically situated players?

And perhaps I should have listed demographers among those experts contributing to policy discussions, for, as I have assumed but not stated explicitly, population is itself a key health-policy variable—framing questions, making issues socially visible, and demanding a response. One thinks of the shift from rural to urban and, especially in the industrial West, the demographic and epidemiological transitions, which—whatever their causes—have produced an older population demanding public-sector responses to ever-increasing levels of chronic and incapacitating disease. (In a similar way, it has been argued, the focus on children in a low-childhood-mortality environment has shaped public-health policies at the local and national levels.) The distribution of potential patients provides not only challenges to the delivery of health care but has, in our advocacy-oriented society, created a powerful pressure group—older citizens who vote.

A parallel argument might be made about globalization, which, as has become clear, threatens to create a social and biological as well as an intellectual, economic, and geopolitical community. AIDS and multi-drug-resistant tuberculosis—and emerging diseases in general—imply and foreshadow a world of risk and interconnection. And that threatening world implies a linked moral and policy dilemma; academic and government researchers and pharmaceutical companies have devoted comparatively little effort to a variety of infectious and parasitic ailments that still kill and disable in the developing world. But as we are well aware, the comparatively recent visibility of the problem has already constituted a novel policy variable.[14]

STRUCTURED CONFLICT

An overarching theme in my argument is the persistence of structured conflicts in American health care, in some ways a consequence of the very complexity and scale of the system and of the cumulative ad hoc decisions that have helped constitute it. These conflicts are at once products, constituents, and predictors of history.[15] I would like to suggest seven such conflicts:

1. The place of the market as resource allocator in a system legitimated historically in rather different terms. Most of us find it difficult to accept instrumental notions of efficiency and market rationality as the ultimate determinant of available health-care options.

2. The problem of valuing outcomes. How does one measure an effective health-care system when there is no easy metric for either clinical efficacy or humane outcome—or ultimately for disaggregating the two?

3. The inherent conflict at several levels between global standards of medicine and local contexts of use.

4. The question of boundaries. Where does medicine stop and something else begin?

5. The immanence and omnipresence of technological and institutional change in a system ill-organized to anticipate and contend with the consequences of such often unpredictable yet culturally valued innovations.

6. The problems associated with decentered and fragmented loci of power, and not only in the obvious realm of government(s).

7. The way in which medicine as a fundamental social function reflects, incorporates, and acts out more general aspects of social hierarchy, status, and power.

Such issues will continue to shape the medical care available to ordinary Americans in the foreseeable future. Let me say a bit more about each of these continuing conflicts.

Medicine has always foraged aggressively in the world even as it has consistently invoked and often acted out a heritage of the sacred and the selfless. In today's highly bureaucratic society, however, the sometimes uneasy relationship between soul and body, between the selfless and the material has been particularly strained by increases in scale and by relatively novel and in some ways inconsistent claims for the market itself as a rationalizing mechanism. The efficacy claimed for the market's discipline fits uneasily with more traditional ideas about healing the sick and defining humane ends.

This is a conflict that has long been implicit but has become increasingly explicit. As the recent history of managed care or of the pharmaceutical industry has made abundantly clear, it is difficult to implement—and effectively sell to the American public—the notion that private-sector competition provides the best available mechanism for achieving a stable and responsive medical environment. Most Americans feel that health care, especially *their* health care, is a right and not a commodity. Indignant reaction to cuts in traditional employer-based health insurance schemes reflect such assumptions as well as a growing disquietude at the increasing number of uninsured generally. But while this sense of moral entitlement constitutes a political reality, it conflicts with another ideological reality: a powerful suspicion of government and an equally widespread assumption that the market is itself a technology that can solve problems. Free market advocates assume, if not always explicitly, that value conflicts are illusory because ultimately the market constitutes the best available mechanism for providing the greatest good to the greatest number.

But this argument, I would suggest, constitutes another example of an endemic ill in our society: a reductionist solution to a holistic problem. The market and its decision-making rationality may be a necessary technology in a complex world, but it is clearly not in itself a sufficient or autonomous one. Market incentives may be powerful, but they are not the only ones. Moreover, political power, as well as institutional interest, can distort and, in part, constitute the market. Hope constitutes another powerful externality. Humanitarian traditions allied with utopian techno-fantasies of life extended and pain forestalled have helped enable an atmosphere of ever-increasing costs and posed the insistent question of how those costs—material and existential—are ultimately to be repaid. The benefits of America's highly technical health-care system are elusive and not easily reduced to consistent measurable terms.

MEASURING EFFICACY

We have neither an easily agreed-upon metric for clinical excellence nor a metric for misery. The extension of the average lifespan and the lowering of infant and maternal mortality are clearly useful and appealingly concrete measures, and they constitute a powerful rhetorical as well as political argument. In an aging society beset by anxieties concerning health and seduced by technologies promising cures, however, such aggregate numbers are clearly insufficient: thus, the creation of concepts such as quality of life or quality-adjusted life years. But such formal constructs inevitably fail to capture and balance the costs and benefits involved in the health-care enterprise—in terms of both individuals and the collective. There is something both dissatisfying yet inevitable about such attempts to provide measurable yardsticks for immeasurable goods.

This is only one part of a much larger difficulty in arriving at consensus in health-care policies, and that is the question of defining success. What is efficacy in health care, and how does it relate to physiological or social function—themselves far from transparent terms? What is to be measured, and against what standard? How does one frame moral and political judgments about the appropriate distribution of resources? How does one implement decisions concerning tradeoffs between quality and cost? But one need hardly go on; these are the familiar complaints of health reformers and a staple of bioethical as well as political angst. No end is in sight.

GLOBAL VERSUS LOCAL

At some level medicine is objective and universal. Most of us think of it that way—as a continually evolving body of accessible knowledge, practices, and tools. Yet we are equally aware that clinical practice varies dramatically from place to place, even within the continental United States, and that not all of these differences can be accounted for by inequality of resources. If we add an international dimension, of course, differences are even more dramatic (one thinks of the Johns Hopkins–trained physician practicing in Lagos or Lima—even if in an elite, well-equipped hospital). Available resources and institutional realities determine the continuing negotiations between the hypothetically possible and the practically achievable, but that tension between the hypothetical best—universal—solution and the implacable contingencies of local circumstance creates a permanent and recurring tension. There is no easy strategy for balancing the generalized truth of the laboratory finding or of the randomized trial against the place- and time-specific context in which that knowledge is applied. Such structured conflicts turn on inevitably contested questions of legitimate authority and the balancing of available resources and social priorities.

BOUNDARY TENSIONS

This balance is only exacerbated—intensified—by another dimension of this question. And that is the matter of boundaries, of defining where medicine ends and everything else begins. How much does the goal of health promotion, for example, trump other values and interests? Are cigarettes drugs? When does a nutritional supplement become a drug appropriate for FDA oversight? Is obesity an issue for clinical medicine or a social, structural, and cultural and, thus, social policy problem? Or is obesity an issue of individual responsibility masquerading as health policy? Is violence a problem for the Centers for Disease Control or for the criminal justice system? Similarly, is substance abuse the responsibility of the psychiatrist, the social worker, the minister, or the policeman? Such boundaries are bitterly contested at many levels and on many battlefields, but the burden remains the same: who is responsible and how much does the need to focus on individual agency trump the determinism implicit in the medicalization of problematic behaviors?

There is also a question of policing. Is the practice of medicine to be internally policed by our contemporary version of guild authorities, the medical societies and specialty boards? Or is medical practice subject in some measure to government and the courts? We have all been made aware of the controversy surrounding what has come to be called "complementary" or "alternative" medicine—and the questions it poses in regard to the institutional boundaries that define the physician's role and the knowledge that legitimates the regular profession's status.[16]

Preventive medicine in general provides another area for debate. Toxic substances, cigarettes, and fetal alcohol syndrome represent occasions for debate and contestation. Is fetal screening for genetic defects a morally neutral technical option—or an occasion for religious introspection, since abortion is so closely linked with screening? How do we balance individual clinical judgments, cost-benefit calculations, and the ever-shifting consensus of evidence-based medicine in justifying particular diagnostic or therapeutic procedures? These are questions of jurisdiction and boundaries, but also occasions for the exercise of social authority. They will continue to remain objects of conflict and contestation.

TECHNOLOGICAL CHANGE AND ITS DISCONTENTS

The only thing more predictable than continued technological change in medicine is the debate and conflict surrounding such change. All those social expectations that encourage and support innovation also nurture conflict over access and specialty control, over economic costs and ethical appropriateness. Humane concerns about the impersonal attributes of high-tech medicine can be traced back to the turn of the twentieth century, when available technologies were, in retrospect, rather crude. Anxieties about cost have an equally long pedigree. More recently, economic and ethical criticisms by most Americans—and not just bioethicists and health economists—have focused on the extraordinary and cost-ineffective technologies involved in preserving life in extremity. The intensive care unit and the respirator represent nemeses that threaten everyone, regardless of social class.[17] I could refer as well to the widely discussed dangers of overdiagnosis and the creation of a legion of asymptomatic—yet therapy-implying—ills. And we live with ever-present dystopian fantasies of human cloning and bodies kept alive with an ever-increasing variety of spare parts.

LOCI OF POWER

The contexts in which medical power is exerted are, as I have implied, fragmented and not always consistent. At one extreme is the bedside—the site of the physician's root legitimacy as healer. Very different sorts of power reside in the intellectual, bureaucratic, and administrative factors that constrain and shape the individual physician in his or her practice. I refer to everything from modes of reimbursement to the practical constraints implicit in disease protocols, evidence-based medicine, and the practice guidelines adopted by large insurers. Physicians have traditionally seen the clinical relationship as theirs to control—citing both ethical and logical reasons to maintain the bedside space as appropriately subject to the individual physician's control. In today's highly technical and increasingly bureaucratized society, that interpersonal decision-making authority has been increasingly compromised by considerations of efficacy, cost-benefit, and all those generalized rationalities that constrain clinical choice. The results of randomized clinical trials, the conclusions of clinical epidemiology, the algorithms, protocols, and thresholds of insurance administrators all imply a structured conflict at the level of decision making—a conflict that remains central not only to what might be called the macro policy of insurance provision or hospital funding, but to the micropolicy of everyday clinical decision making. There is no easy strategy for balancing the generalized truth of the laboratory finding or of the randomized trial against the site and individual-physician-centered truth legitimated by the patient's particular biological and social circumstances. That conflict between the generalized truth and its specific application will not be solved by some meta-consensus committee or cost-benefit analysis.

The question of appropriate authority in clinical decision making is a question of power and politics—macro and micro—but it is also a question of ethics, political theory, and policy. In America's complex federal system there are some obvious linkage questions. Where is the appropriate balance between state, federal, and local authority? It is a recurrent issue in American history, manifested now in struggles over such visible issues as abortion policy or Oregon's "right to die" legislation or California's medical marijuana statutes. Even more fundamental is the conflict between these traditional definitions of local jurisdiction and the instinctive moral feeling that health care should be universally accessible—and not determined by one's place of residence along with one's income. Who is to set priorities in everything from the distribution of research support to the provision of funds to subvent care for the poor and uninsured? Tax policy, as well as randomized clinical trials, helps shape the care delivered ultimately to particular patients. The political and administrative realities of health care also demonstrate the complexity of the interactions and interdependencies among local, state, and federal levels of need and responsibility. American federalism implies local differences in social need and available resources—intractable realities that often seem inconsistent with the legitimating vision of medicine as an objective—and thus universal—body of knowledge and practice.[18]

Finally, of course, medicine mirrors all the ambiguities and conflicts that characterize and mark our society: welfare policy, specific attitudes toward race as in drug laws and aid to dependent children, attitudes toward sexuality as manifested in AIDS policy and sex education. Changing media and communication realities imply changing social relationships; one need only watch consumer advertising on network television to extrapolate a new world of chronic disease, an aging population, and an ongoing set of political filiations and agendas. The same might be said of the ability of the Internet to crystallize and distribute the message of disease advocacy groups, for example, as much as political parties themselves. And with an aging population and the increasing economic prominence of the health-care enterprise, it was inevitable that health issues would be as subject to the same lobbying and tactical political considerations as tax or industrial policy.

HISTORY AND STRUCTURED CONTINGENCY

Policy is always history. Events in the past define the possible and the desirable, set tasks, and define rewards, viable choices, and thus the range of possible outcomes. As we move through time those choices reconfigure themselves, and trends may establish themselves—but at any given point the available options are highly structured. It is the historian's disciplinary task to discern those likelihoods. Most important, what history can and should contribute to the world of policy and politics is its fundamental sense of context and complexity, of the determined and the negotiated. The setting of policy is contingent, but it is a structured contingency.[19] There should in this sense be no gap between history and policy any more than there is between any of the other social sciences and the making and administration of policy.

But there is. In some ways it is a problem of audience and expectation. Historians make uncomfortable prognosticators. We feel that historians should look backward, while the essence of policy is to look forward—as though the past is not in the present and the present in the future. History cannot predict what will happen; it is a more useful tool for predicting what will not happen. Or, to put it another way, defining nonchoices is an important way of thinking about choices. But history is not simply a database from which policy makers and policy scientists can mine rhetorically useful bits of admonition or encouragement; history is a complex discipline with its own constraints and necessities—its own substantive consensus.

Every participant in the world of policy and practice has his or her own history that creates community and legitimates policy choices. Even the work of academic historians is inevitably a source of decontextualized data for real-world actors who deploy it in the context of their particular visions of policy. But the historian's primary context is the world of other historians, and it is this very distance from the policy arena that makes the historian's perspective so valuable. To be effective historians must maintain their disciplinary identity, their own criteria of achievement and canons of excellence. We are spectators at the policy dogfight—and may even lay the odd wager—but at least we do not own any of the combatants in the pit.

NOTES

1. In our disciplined and bureaucratic society, it also assumes a role for experts (including lobbyists for a variety of interests) interacting with decision makers—both elected and unelected—in what is sometimes called a policy community. Policy may also imply claims to legitimacy based on presumed connections among data, data analysis, and subsequent decisions and practices. Finally, at the national level, domestic policy presumes a logic of center and periphery, of centralized decision making, even if particular decisions mandate the devolution of implementation to the periphery (as in Medicaid, for example, or, in its rather different sphere, specialty credentialing). Even the term *health policy* implies too neatly unified a sphere of action, as though health practices and expenditures could be insulated in a complex and interactive world of political and economic decision making. On specialism, see Rosemary Stevens, *American Medicine and the Public Interest* (Berkeley: University of California Press, 1998) and "Medical Specialization as American Health Policy: Interweaving Public and Private Roles," in *History and Health Policy in the United States: Putting the Past Back in,* ed. Rosemary A. Stevens, Charles E. Rosenberg, and Lawton R. Burns (New Brunswick, N.J.: Rutgers University Press, 2006).

2. Which is not to say that such contributions have not and will not continue to have their uses, sometimes central ones, but only in appropriate contexts of power and advocacy.

3. The de facto disdain for history acted out by many policymakers is itself an historical artifact that demands explanation in historical and cultural, not logical, terms. Yet every discipline and every political position has its own proprietary history: the nurse's history is different from the physician's, the free market from that understood by advocates of a robust government role in health care, but one thing all such actors share is the willingness to frame and legitimate policy commitments in tactically convenient historical terms.

4. For an example of the use of history to reflect on public policy, see Daniel M. Fox, *Health Policies, Health Politics: The British and American Experience, 1911–1965* (Princeton, N.J.: Princeton University Press, 1986) and *Power and Illness: The Failure and Future of American Health Policy* (Berkeley: University of California Press, 1993).

5. On hospitals, see Charles E. Rosenberg, *The Care of Strangers: The Rise of America's Hospital System* (New York: Basic Books 1987), and Rosemary A. Stevens, *In Sickness and in Wealth: American Hospitals in the Twentieth Century,* rev. ed. (Baltimore: Johns Hopkins University Press, 1999 [1989]).

6. Although the profession is characterized by a historically negotiated self-governing status, that autonomy is itself legitimated in part by medicine's traditional identification with the public interest.

7. There were some other areas in which the federal government did begin to support scientific investigation in the nineteenth century—most prominently agriculture and then engineering, the geological survey, and (indirectly) through the support of higher education generally.

8. It should be emphasized that American government was at every level erratic and inconsistent in performing these functions.

9. It is not without significance that we conventionally refer to our "health-care" system, not our "medical-care" system, reflecting another usage distinguishing public from private—and thus obscuring the disparate strands that collectively make up these conventional categories; even the "private" practice of medicine has included aspects of what might be called "public" medicine—as in immunizations against infectious disease.

10. There is a countervailing—dystopian or, one might call, it Frankenstein-ian—vision of the increasing dangers implicit in technological innovation, but I would argue that such anxieties have always been a minority theme.

11. Which is not to say that such relationships are unprecedented, but that they are often *perceived* as somehow radically novel—and morally compromising.

12. See "The New Enchantment: Genetics, Medicine, and Society," chapter 6 in this volume.

13. I refer to everything from hospitals and outpatient clinics to specialty boards and health maintenance organizations.

14. It should be noted that we live to some extent in an increasingly global world of medical personnel and medical education as well as of finance and manufacturing.

15. Most of these conflicts are structured into Western medicine generally and not limited to North America. My argument focuses, however, on the United States.

16. See "Alternative to What? Complementary to Whom? On the Scientific Project in Medicine," chapter 7 in this volume.

17. For a defense of the economic and human rationality of continued investment in high-tech medicine, see David Cutler, *Your Money or Your Life: Strong Medicine for America's Health Care System* (New York: Oxford University Press, 2004).

18. It is striking how studies that demonstrate regional patterns of inconsistency in clinical practice simply presume that the inconsistency speaks for itself—both logically and morally—as indicating irrationality and, thus, need for change.

19. Paul David's term *path dependency* has become commonplace in the policy world; to a historian it is in operational terms another word for history—and a de facto substitute for it. I have used the term *structured contingency* because I had hoped to emphasize that key area of contention between the determined and the negotiated—the structured as opposed to the contingent; see David, "Clio and the Economics of QWERTY," *American Economic Review* 75 (1985): 332–37, and "Understanding the Economics of QWERTY: The Necessity of History," in *Economic History and the Modern Economist,* ed. William N. Parker (Oxford: Basil Blackwell, 1986); W. Brian Arthur, *Increasing Returns and Path Dependence in the Economy* (Ann Arbor: University of Michigan Press, 1994); and Jack A. Goldstone, "Initial Conditions, General Laws, Path Dependence, and Explanation in Historical Sociology," *American Journal of Sociology* 194 (1998): 829–45.

ACKNOWLEDGMENTS

— —

THIS BOOK IS BASED ON MY WORK AS A professional historian—and as an observer living through an era in which medicine moved from the margins of everyday public discourse to the center. Stories of medical discovery, health policy politics, and ethical (and often political) controversy appear regularly in the news pages, while accounts of drug company profits, of corporate lobbying, of hospital mergers and insurance practices, of personal bankruptcy and malpractice saber rattling may lead the business pages. And, of course, television and the Internet mirror, intensify, and in part constitute our particular world of medical hopes and anxieties. This book is inspired by such public debate but is not addressed to specific policy choices: how we should tweak health insurance or cap malpractice awards. It seeks instead to think critically about fundamental aspects of medicine ultimately but complexly related to such immediate issues—how we think about disease, how we think about the moral and intellectual responsibilities of the profession, how we conceive of medicine as both inhabiting the market yet by its nature transcending it. It is my hope in this book to not only think about medicine but also to suggest how we can think with medicine—to think about the society that supports and in part constitutes the world of sanctioned healing.

I have benefited from so many conversations and suggestions that it is hard to recount or specify all those friends and fellow students of medicine's history and social role who have contributed to this book, but let me mention a few whose comments over the years have been consistently helpful: Robert Aronowitz, Allan Brandt, Gretchen Condran, Leon Eisenberg, Drew Faust, Steven Feierman, Renee Fox, Susan Lindee, David Mechanic, and Barbara Rosenkrantz. I have also learned a great deal from my graduate students at the University of Pennsylvania and at Harvard, who have taught me at least as much as I taught them.

I would also like to thank the following journals and publishers for permission to reprint (with minor changes) material that originally appeared in their publications. Chapter 2 first appeared as "The Tyranny of Diagnosis: Specific Entities and Individual Experience," in the *Milbank Quarterly* 80 (2002): 237–60; chapters 3 and 4 appeared in *Perspectives in Biology and Medicine:* "Contested Boundaries: Psychiatry, Disease, and Diagnosis," 49 (2006): 407–24, and "Banishing Risk: Or the More Things Change the More They Remain the Same," 39 (1995): 175–97; chapter 5 appeared as "Pathologies of Progress: The Idea of Civilization as Risk" in the *Bulletin of the History of Medicine* 72 (1998): 714–30; "Holism in Twentieth-Century Medicine," in *Greater than the Parts: Holism in Biomedicine, 1920–1950* (New York: Oxford University Press, 1998), pp. 335–55, appears here as chapter 8. Chapters 9 and 10 were originally published in slightly different form as "Meanings, Policies, and Medicine: On the Bioethical Enterprise and History," *Daedalus* 128 (1999): 27–46, and as "Anticipated Consequences: Historians, History, and Health Policy," in *History and Health Policy in the United States: Putting the Past Back In,* ed. Rosemary A. Stevens, Charles E. Rosenberg, and Lawton R. Burns (New Brunswick, N.J.: Rutgers University Press, 2006), pp. 13–31, respectively.

INDEX

Access to health care, 2, 201
Acute care–oriented medicine, 153–54
Adaptation, 86, 91
Addictions, 40
Addictive personality, 70
Advertising in health care, 167, 170
Aging population, 2, 7
AIDS, 88, 152, 153, 154–55, 156, 192,
  196, 202
Alcoholism, 63, 64, 66
Alternative and complementary
  medicine, 113–32; boundaries
  between Western medicine and,
  114–16, 199; clinical trials of, 127;
  debate about, 113–14; disdain of
  regular medicine toward, 124–25;
  emergence of, 122–25; holism and,
  152; in managing diversity, 125–27;
  medical tradition and, 116–20;
  sources of patronage for, 127–30;
  tolerance of, 125–26; viewing
  mainstream medicine in light of,
  130–32

American Institute for Preventive
  Medicine, 72
American Medical Association, 124,
  125
Antibiotics, 4, 24, 87–88, 153
Autonomy, 174, 177–78, 180

Beard, George M., 28, 45–47, 78, 83, 84
Beecher, Henry, 177
Beneficence, 178, 180
Bioethics, 153, 166–82; activities of,
  178; codes of medical ethics, 8;
  historical sensibility and, 179–82; as
  historical subject, 176–79; history,
  contingency and, 182; medical care
  and social obligation, 167–69;
  medicine and meanings, 170–72;
  novel realities and, 172–75; principles
  of, 177–78; research and, 180; roles
  and sites of social action, 178–79;
  as social technology, 167; technical
  capacity and, 200
Biomedicine, 116, 130

Biopsychosocial approach, 9, 150
Bioterrorism, 2
Blue Cross and Blue Shield, 4
Body mass index, 31
Body-mind holism, 128–29, 152
Boundary tensions, 199–200; in psychiatric diagnosis, 45–48, 53; between Western and alternative medicine, 114–16, 199
Bureaucratization, 5, 8, 13, 25–26, 27, 31, 201
Burnet, Macfarlane, 88

Cabot, Richard, 23
Cancer, 73, 88, 89, 168
Case records, 21–22
Center for Complementary and Alternative Medicine, 113, 127
Centers for Disease Control, 199
Cheyne, George, 81
Cholesterol level, 67, 70
Chronic disease, 62–63, 67–70, 77–78; heredity and, 102; management of, 15, 67, 96, 154
Civilization and disease, 77–92; mental illness, 80–81, 82–83, 84; persistent meanings, 89–92; social change and sensory overload, 82–85; Stone Age man in the fast lane, 77, 85–89; tradition of right living, 79–82
Clinical decision making, 6–7, 97, 174; authority in, 200–201
Clinical trials, 201; of alternative therapies, 127
Codes of medical ethics, 8
Communication, 83–84, 106
Computerization, 26
Constitutional factors, 101–4, 106, 107
Credentialing, 123, 180
Crile, George, 85–86
Criminal responsibility, 38–40
Cullen, William, 17
Cultural competence, 128

de Kruif, Paul, 181
Diabetes, 2, 24, 67, 96
Diagnosis, 13–34; psychiatric, 14, 26, 33, 41; social function of, 15–16, 32–34
Diagnosis-related groups, 15, 157
Diagnostic and Statistical Manual of Mental Disorders, 14, 26, 33, 41, 50, 157
Dietary factors, 129, 156
Differential diagnosis, 23
Disease: acute vs. chronic, 62–63; banishing risk for, 60–74; causation of, 3, 60, 72, 78–79; civilization and, 77–92; criminal responsibility and, 38–40; diagnosis of, 13–34; evolution of concepts of, 3, 6, 13–15, 60–62, 64; genetic, 96–110; vs. illness, 34, 151; lifestyle and, 60–62, 67, 69, 74, 77–79, 88–89, 129; management of, 15, 67, 96, 154; psychiatric, 38–56; response to therapeutics, 25; as social entity, 5–6, 24–27
Disease specificity, 16–23, 157; historical lack of, 61; mechanism and, 64–68; paradoxes of, 27–32; for psychiatric disorders, 42–45
Disparities in health care access, 2, 201
DNA era, 100
Domestic medicine, 120–21
Drugs, 4, 6, 87, 153; governmental regulation of, 188; marketing of, 2, 6, 32; for psychiatric diagnoses, 50–51
Dubos, René, 148

Ecological holism, 146–47
Efficacy of health care, 198
Eisenberg, David M., 125
Emerging diseases, 2, 88, 195
Engel, George, 150
Engels, Friedrich, 146
Entitlement to health care, 197

Environmental factors and disease: civilization and, 77–82, 85–89; constitution and, 102–3

Epidemics, 2, 3, 18, 78

Epidemiology, 20–21, 24, 88–89

Erichsen, John, 46–47

Euphoria of magic rationalism, 97

Euthanasia, 171

Evidence-based medicine, 6, 7, 15, 24, 26, 200

Expectations: entitlement to health care, 197; inconsistent, 1–2

Faber, Knud, 17

Fetal alcohol syndrome, 51–52, 199

Fisher, R. A., 100

Forensic psychiatry, 28, 38–39

Freud, Sigmund, 92

Funding for disease-targeted research, lobbying for, 41, 168

Galen, 61

Garrod, A. B., 100

Gene therapy, 96, 110, 191

Genetic disease, 16, 96–110; acquisition of knowledge about, 100, 108; constitution and, 101–4, 106, 107; counseling about, 99; historical notions of, 99–106; "Jewish," 98–99; mental illness as, 173; screening for, 99, 200; social impact of knowledge about, 108–10; testing for, 30–31

Germ theory, 3, 19, 66, 141, 145

Global health care, 195–96, 198–99

Goldstein, Andrew, 38

Governmental roles, 2, 4, 5, 174, 188–89, 199; federal, state, and local authority, 201

Health care costs, 2, 7; containment of, 15; efficacy and, 198; third-party payment for, 4–5, 24

Health-care system, 167, 169, 170; disparities in access to, 2, 201; holism and, 155; as part of political system, 193–95; principled criticism of, 177; structured conflicts in, 196–203; values and structures of, 190–93

Health insurance, 2, 4–5, 24, 43, 197

Health policy, 185–203; cultural politics of, 190–91; "formation" of, 194–95; globalization and, 195–96; process, system, and structure of, 193–96; public and private, 186–90; structured conflicts and, 196–203; values and structures, 190–93

Health promotion, 199

Healthy People 2000, 68–69

Herbal remedies, 126–27

Heredity. *See* Genetic disease

Historical holism, 142–44

Historical notions of disease, 3–6, 13–15; hereditary diseases, 99–106; psychiatric disorders, 44–48

History: of alternative medicine, 122–25; of American medicine, 122–23; bioethics and, 176–82; health policy and, 185–203, 202–3; of traditional medicine, 116–20

Holism in medicine, 70–71, 118, 139–58; alternative medicine and, 152; categories of, 142; ecological, 146–47; evolution of, 141–42; health-care system and, 155; historical, 142–44; meaning of, 139–40; new, 150–51, 153–55; nursing and, 151; organismic, 144–46; paradoxes of, 155–58; psychiatry and, 151–52; vs. reductionism, 140–41, 149, 153, 155–58; social medicine and, 152–53; social position of articulators of, 149–53; worldview, 147–49

Homeopathy, 121, 152

Homeostasis, 91

Homosexuality, 14, 41, 46, 47, 171
Hospitals, 3–4, 21, 25, 167–68
Hypertension, 67, 70, 88
Hypochondria, 46
Hysteria, 78

Illich, Ivan, 148
Illness vs. disease, 34, 151
Inconsistent expectations, 7–8
Infectious disease, 3, 18, 62–63, 66,
    191; conquest of, 86, 87, 154;
    emerging, 2, 88, 195; germ theory
    of, 3, 19, 66, 141, 145; globalization
    and, 195–96; urban stress and, 78
Information technology, 5–6
Informed consent, 178, 180, 181
Innovations in medicine, 2, 24, 200.
    See also Technology
Institutional review boards, 178, 180
Instruments of precision, 19–20
Insulin, 4, 24, 87

Justice, 174, 178

Kaiser Plan, 5
Kaptchuk, Ted J., 125
Kaysen, Susanna, 50

Laing, R. D., 50
Lay healers, 118–19, 120–21
Legal issues: drug regulation, 188;
    mental illness and criminal
    responsibility, 38–40
Lewis, Sinclair, 191
Life expectancy, 2, 66, 154, 198
Lifestyle and disease, 60–62, 67, 69, 74,
    77–79, 88–89, 129; personal responsibility,
    192; tradition of right living,
    79–82
Lobbying for disease-targeted research
    funding, 41, 168
Lombroso, Cesare, 28
Lyme disease, 71, 128

Managed care, 5, 29, 174, 180, 197
Market and health care, 167, 170, 189,
    197–98
McKeown, Thomas, 86–87, 146
Mechanisms of disease, 173; meanings
    and, 72–74, 129–30; specificity and,
    64–68
Medical citizenship, 9, 10–11
Medicalization of deviance, 28, 39, 42,
    199
Medicare and Medicaid, 4, 188
Medicine and meanings, 170–72
Medicine as bounded profession,
    120–24
Medicine as mirror of society, 202
Mendel, Gregor, 100
Mengele, Josef, 177
Mental illness. See Psychiatric diagnoses
Mind-body relationships, 128–29, 152
Morality, 7, 10; bioethics and, 174–76;
    disease and, 61, 62–64, 65, 71, 171;
    entitlement to health care, 197
Morgan, T. H., 100
Musser, John H., 23

National Cancer Institute, 168
National Institutes of Health, 113, 127,
    168
Neo-Hippocraticism, 144
Neurasthenia, 28, 45–47, 78, 83
Nightingale, Florence, 151
Nonmaleficence, 178
Nordau, Max, 83–84
Nosology, 17, 21, 23, 157;
    administrative role of, 31;
    iatrogenesis of, 30; psychiatric,
    14, 26, 33, 41, 50, 157
Nursing holism, 151

Obesity/overweight, 31–32, 54, 107,
    199
Organismic holism, 144–46
"Overstress," 47

Pathogenic consequences of progress, 77–92

Patients, 1, 6; access to health care, 2, 201; beliefs of entitlement, 197; inequities among, 174

Pauling, Linus, 100

Pellagra, 156

Personality factors, 70

Pharmaceutical industry, 2, 6, 32, 41, 51, 55, 197

Physician-patient relationship, 9–11, 68, 107–8, 145

Physicians, 174; as bounded profession, 120–24; historical legitimacy of, 117–20, 190; lay healers and, 118–19, 120–21; loci of power, 200–201; medical citizenship of, 10–11; policy and responsibilities of, 8; respect for, 2; social status of, 66, 190

Physicians for Social Responsibility, 152

Pinel, Philippe, 49

Practice guidelines, 6–7, 15, 200

Predisposition to disease, 63, 65, 101–2

Preventive medicine, 199–200

Progress: holism and, 153–55; pathogenic consequences of, 77–92

Protodisease states, 68, 70

Psychiatric diagnoses, 14–15, 22, 26, 28, 33, 38–56; arbitrary nature of, 50; categories of, 14, 26, 33, 41–42; civilization and, 80–81, 82–83, 84; coding of, 54; conflict and continuity of, 53–56; criminal responsibility and, 38–40; culture and, 55–56; drugs and, 50–51; expanding boundaries of, 45–48, 53; genetics of, 173; historical concepts of, 44–48; insurance coverage for, 43; medicalization of deviance, 28, 39, 42, 199; nosology of, 14, 26, 33, 41, 50; reductionist explanations of, 39, 44, 48, 49–50, 53–54; somaticization

and legitimacy of, 46–53; specificity of, 42–45

Psychiatric holism, 151–52

Psychosomatic medicine, 70–71, 151–52

Public-health holism, 152–53

"Quackery," 124

Quality assurance, 29

Quality of life, 198

Railroad spine, 44, 46–47

Ray, Isaac, 83

Reductionism, 4, 9, 19, 28, 34, 39, 44, 48, 49–50, 53–54, 65, 73, 104, 192; vs. holism, 140–41, 149, 153, 155–58

Reimbursement, 4–5, 24, 54

Religion, 63–64

Research: archival, 14; bioethics and, 181; genetic, 96–110, 169; health policy, 186; lobbying for funding for, 41, 168

Richardson, Benjamin, 81

Risk for disease: banishing of, 60–74; civilization as, 77–92; lifestyle and, 60–62, 67, 69, 74, 77–82, 88–89, 129, 192

Roosevelt, Franklin, 193

Rush, Benjamin, 46, 49, 81

Scientific medicine, 7, 20, 24, 30, 66, 73, 119; boundaries between alternative medicine and, 114–16; faith in, 191; normal fallibility of, 115; origins of Western medicine, 116–17

Self-denial, 69

Selye, Hans, 86

Sickle-cell anemia, 98

Sigerist, Henry, 146

Social change and sensory overload, 82–85

Social diversity, 174–75

Social efficacy, 9–10
Social function of disease, 15–16, 32–34
Social medicine, 8–9, 87, 104; genetic research and, 108–10; holism and, 152–53
Society and medicine, 2–9
Specialism, 123, 180
Stem cell research, 97, 109, 191
Stigmatization, 28, 29, 44, 47, 55, 73, 170
Stone Age man in the fast lane, 77, 85–89
Stress, 47, 91; urban life and, 77, 78, 82, 89
Structured conflicts in health care, 196–203; boundary tensions, 199–200; efficacy, 198; global vs. local, 198–99; history and structured contingency, 202–3; loci of power, 200–202; the market, 197–98; medicine as mirror of society, 202; technological change, 200
Substance abuse, 40, 63, 64, 66, 199
Szasz, Thomas, 50

Tay-Sachs disease, 98–99
Technology, 1–2, 7, 8, 13, 21, 24, 33, 197–98; bioethics and human implications of, 167, 168–69; conflicts surrounding changes in, 200; social change, sensory overload and, 82–85
Temkin, Owsei, 131
Third-party payment, 4–5
Traditional medicine, 116–20
Trotter, Thomas, 80
Trust, 2
Tuke, D. Hack, 80–81
Type-A personality, 70

Urban stress, 77, 78, 82, 89
U.S. Department of Health and Human Services, 68

Values, 166, 174, 190–93
Veterans' Administration, 4, 188
Virchow, Rudolf, 146

Wade, Nicholas, 49
Weissmann, August, 100
Western medicine: boundaries between alternative medicine and, 114–16, 199; holism and, 139–58; nature of problems confronting, 166; origins of, 116–17
Worldview holism, 147–49
Wright, Sewall, 100

Yates, Andrea, 38–39